Jungle
Gems

Map of the Jungle

The Great Mountains

The Labyrinth of Roots

The Waterfall

The Clearing

The Jungle River

The Village

The Museum

The Lowlands

N

Jungle Gems

A NATURALIST'S TALE

YIKAI ZHANG

unbound

This edition first published in Great Britain 2019

Unbound

6th Floor Mutual House, 70 Conduit Street, London W1S 2GF

www.unbound.com

Text Design by PDQ

A CIP record for this book is available from the British Library

ISBN 978-1-78352-792-2 (trade hbk)

ISBN 978-1-78352-794-6 (ebook)

Printed in Slovenia by DZS Grafik

1   3   5   7   9   8   6   4   2

Max Barclay,
who ignited my passion for entomology;

Gabriel Forbes-Sempill,
who introduced me to the wonderful world of art;

and

my long-suffering family,
for supporting my hobby and, at times,
putting up with a home full of insects.

CONTENTS

FOREWORD

It is 250 years since Captain Cook led the *Endeavour* expedition to the South Pacific, ostensibly to observe the transit of the planet Venus across the sun, a phenomenon apparently best seen from the then newly mapped island paradise of Tahiti. Cook was accompanied on this voyage by a naturalist, the headstrong Joseph Banks, who brought back hundreds of botanical and zoological treasures. Many of these, from Australia and New Zealand, were the first specimens of their kind to be systematically collected, and most remain at the Natural History Museum in London. These collections are a kind of time capsule, a window into the past, providing an insight into these environments before two and a half centuries of 'progress' had rolled over them.

The voyage of Cook and Banks is often described as the beginning of a new age of scientific discovery, when the furthering of human knowledge, rather than the fleeting dreams of gold, land and empire, became a primary objective. This age of curiosity-driven exploration continued throughout the nineteenth century, and as Yikai's highly entertaining, thought-provoking and beautifully illustrated book shows, it is far from over today. With the destruction humans have wrought on the natural environment, discovery and documentation of nature have acquired a new urgency, and the great museums that collate and distil the fruits of these voyages are becoming ever more important as archives of the world's biodiversity.

Jungle Gems is a natural history adventure in the classic mould, but with a distinctly modern twist. Its hero sets out in search of beetles, arguably the most diverse group of organisms on the planet, with more than 400,000 species so far named by science. Experience has shown us that a skilled collector and observer, working with such a vast and seemingly infinitely variable group, may see patterns in the natural world that are hidden from others. It is no coincidence that Charles Darwin and Alfred Russel Wallace, the great pioneers of evolutionary thought, were first and foremost enthusiasts of beetle collecting. Wallace stated that 'he who has never observed and studied beetles, passes over more wonders in every field and every copse than the ordinary traveller sees who goes around the world', and it was the depth of observation and intensity of passion of these two amateur beetle collectors, combined with the diversity of their quarry, that led to one of the greatest leaps forward in the history of human thought. As we invent new techniques and new technologies, it is important to remember that there is still no substitute for patient and systematic expert observation and collecting, and that this work remains relevant and plays a key role in how we understand and conserve the natural world.

Yikai's book shows that exploration and curation are not only about observation, but may include a large chunk of detective work as well, and he describes the thrill of discovery as pieces of a puzzle, natural or otherwise, 'drop into place'. It is also splendidly illustrated with the author's own watercolours, which invoke a golden age of natural history illustration.

I hope that *Jungle Gems* will help to inspire a resurgence of explorer naturalists to go out into natural habitats and study and collect natural objects. There is much that we still don't know about the natural world and its component species, and remarkable new discoveries certainly still await the patient, the curious and the intrepid in the jungles and forests, and indeed in the copses and fields, of the world.

Maxwell V. L. Barclay
Senior Curator of Coleoptera, Natural History Museum,
London
August 2018

PREFACE

It is with great fondness that I look back over the past year of my life, and realise just how much I have matured in tandem with my own creation. Every writer knows the struggle of finding the right words to express their feelings, and every artist knows the struggle of being confronted by a blank canvas before they even get the chance to appreciate their last piece. I have experienced my fair share of both, and then some more to compensate for my lack of formal training in art and literature. Despite the process turning out to be far more difficult than I could possibly have anticipated, I now have the privilege of sitting down in front of the finished work and reminiscing about the good times – the weeks of skipping university lectures to make the first illustrations, the intrepid adventures through the steamy jungles and the many cold winter nights spent at the keyboard, where all my best experiences were processed and distilled into the work before you.

A question I often get asked is 'why insects?', and to be honest it's quite complicated. The simple answer is that they're incredibly diverse and colourful and I find them fascinating, but you'll get that answer from anyone who claims to be interested in insects. Besides being a subject matter dear to my heart, I believe that thinking about the lives of insects (and indeed nature in general) encourages us to be more introspective about our own. One can't help but begin to ponder life's great mysteries when we watch generations of these tiny creatures rise and fall before us like the evening tide. What is the purpose of this cycle of life and death, or does it simply 'exist' without any inherent meaning? Do we really have free will, or are we just glorified machines built by our genetic blueprints, doomed to be slaves of our primitive desires? What if we were told that we had only a few months to live and fulfil our wishes? Would we still be making the same choices as we are making now? Questions such as these are very pertinent, since we are just as mortal as short-lived insects, yet we are somehow oblivious to our mortality unless our lives are in immediate danger. The great irony is that we see insects as simple-minded beings to be trampled on and ignored, yet with all our supposed ingenuity we are arguably more clueless: a colony of ants will work together tirelessly to build an intricate nest with no project leader, no complaints and no quarrels; a human building, on the other hand, would take months of careful planning and negotiation to even begin construction, and imagine the chaos that would ensue should there be any fault in technology or communication! In a way, insects represent purer lifeforms who are not burdened by the worry of forethought and are guided mostly by instinct – their lives are fleeting, brutal and raw, bridging the apparently incompatible spheres of the elemental and sentimental.

Insects are the details in the wonderful world we live in. It is this appreciation of detail, coupled with an insatiable curiosity, that have come to define me as a person. I cannot imagine where I would be if I hadn't been introduced to nature at a young age – or even worse, if I had conformed to my peer group and given in to the ridicule I was often subjected to. Indeed, on looking back over my personal journey, my fascination for insects and the natural world has permeated virtually every aspect of my life and in turn inspired countless positive changes: art stemmed naturally from the patient observations, I took up carpentry in order to build cabinets for my ever-increasing beetle collection, and my love for callisthenics began when I climbed trees to catch the longhorn beetles perched atop the branches. Nature has taught me patience, focus and tenacity, to be grateful for the life I have been given and to explore the enormous potential that lies within every one of us. Although I decided to use insects to convey my story, this book could have been about anything in the natural world: the crustacean community in a rockpool, the seeds of deciduous trees in a hedgerow or the minerals in a drop of pond water. The point is that there is often more to these 'trifles' than meets the eye, and that they contain enough wonder and complexity for a whole lifetime of study. If there is only one thing you take away from this story, I hope that it will be an appreciation for these fascinating details, and a more balanced view of the natural world in general.

You may have picked up this book because of the pretty pictures, or perhaps you just wanted to find out a bit more about insects. Whatever the reason, I hope that by the end of the story you will have found something that inspires you, whether it's to revive your passion for outdoor exploration or to pick up that dusty paintbrush again. But above all, I hope you enjoy the book!

Yikai Zhang
September 2018

AUTHOR'S NOTE

As there is various technical terminology peppered throughout the book, I thought it would be appropriate to introduce the reader to some of the more important ones for a better understanding of the story. I shall first introduce myself as an amateur entomologist, someone who spends a big portion of his spare time collecting and studying insects. My primary interest is in the order Coleoptera, which are commonly known as the beetles, and so an entomologist specialising in this taxonomic group may also be referred to as a 'coleopterist'. If an entomologist is lucky enough to find a yet undescribed species of insect, then a specimen known as the 'holotype' will be used as the standard reference for its naming and description. The holotype of a species will have a red label affixed to distinguish it from the others, as it is considered to be of great scientific importance.

When I received a box of specimens from Hakomoto, an insect dealer from Japan with whom I had professional dealings, I noticed that there was a specimen of *Carabus* collected from the island of Hainan, where it had never been known to exist. Strictly speaking *Carabus* is a genus name comprising over 1,000 species of ground beetles, but is here used to refer to the species represented by Hakomoto's specimen. Usually the name of any biological species would consist of two parts: the genus name and the species name. *Carabus* is the genus name that is common to all the species within the genus, and a further name would be required to specify a certain species. Examples would be the species *Carabus ignimitella* and *Carabus pustulifer*, which literally mean 'fire headdress *Carabus*' and 'blister-carrying *Carabus*' respectively. *Lucanus* and *Rhomborhina* are two other genus names used for a similar effect in the narrative, where the exact species is unspecified or simply irrelevant to the story. Any species from the genus *Carabus* found in Hainan, however, would likely be undescribed and require further molecular tests to ascertain its status – so this enigmatic group naturally became my main quarry throughout the adventure.

The interesting thing about the *Carabus* genus is that they don't seem to be able to survive in tropical climates – unlike many other insects which thrive in moist equatorial forests, these beetles are virtually absent from the tropics and their distribution is confined to the temperate and subtropical parts of the Northern Hemisphere. In the UK there are about a dozen recorded *Carabus* species, and most of them can be found right down at sea level. The only occasions where *Carabus* have been found in the tropics are high up in the mountains, where the altitude creates a temperate climate that is cool enough for the beetles to survive.

It was with this mission that I set off for the mountains of Hainan, in an attempt to discover a species of *Carabus* new to science.

CARABUS

London, June 2017

I took a deep breath, and stepped inside the storage room. The museum curator had his back towards me and was hunched over an old drawer of specimens, apparently unaffected by the dim lighting and suffocating stench of mothballs.

'Max…'

My voice trailed off into silence as there was an air of extreme concentration to his demeanour. He delicately lifted up a tiny speck with his watchmaker forceps, and hovered it under the microscope. I looked over his shoulder to see what on earth could have fascinated him so much, and saw that his focus was justified – those beetles were devilish! With huge alien eyes and curving jaws shaped like sickles, they could have been miniatures from a horror movie.

I immediately recognised them. 'Hey, that's—'

He put up a finger, signalling me to be silent. I could see his hand quivering as the antenna moved closer to its owner, closer to that tiny bead of glue covering the point of fracture. Then, with a calculated thrust of the forceps, the fragment slipped back into the ball-and-socket joint, firmly anchored in place by the cup and glue. The entomologist gave a sigh of satisfaction, stretched his arms into the air and spun round in his swivel chair. If it weren't for the 10 million other dead beetles in this room, one might have thought that this was the only one left in the world.

'Fine holotype of *Dorysthenes walkeri*, that was,' he muttered. 'Shame one of its antennae got knocked off during the relocation.' Max noticed me over the top of his spectacles, and resumed his usual friendly manner. 'Oh, hallo, Yikai! Where did you come from?'

'I'm afraid, Max, that I have failed. I thought I found a *Carabus*, but it sprayed acid into my eyes and escaped.' I handed the specimens to him, and began to apologise breathlessly about my incompetence.

'It is as I expected, then,' he said slowly, not even looking up from the mass of tiny beetles crammed in the resealable packets, which were clearly of greater fascination to him than my failure. 'Indeed, you have done remarkably well. There are lots of interesting Prionocerids in this material which would be better off in Michael's hands.'

'*As expected*?' I gritted my teeth, bitter at the fact that all my hard work had come to nothing. 'But I was so close! I tried pitfall trapping in the higher altitudes and went torching virtually every night. If only I had a couple more weeks I would've got the rascal, but I had to come back for that stupid...'

I continued rambling about *Carabus*, and how impossible it was to catch, without paying any heed to Max. He looked up at me with a mysterious smile, and opened a small cardboard box on the table. I was greeted once more by the sight of those glistening beetles, and the enigmatic specimen that had eluded me throughout my adventures.

'You see, I thought the thing which sprayed me up near the peaks was some sort of *Carabus*.' I pointed at a slender, sinister-looking beetle in the corner. 'But it looked nothing like Hakomoto's one. Much smaller, emerald green and brilliantly iridescent – must've been a whole new species altogether. I later saw a pair of specimens in this abandoned museum, so I'm sure there's a decent population up there.'

'I highly doubt it; that sounds more like *Catascopus* to me,' he mused. 'They're also capable of spraying formic acid, you know?'

I felt like a great weight had just been lifted off my mind, and Max's general demeanour somewhat calmed my frustration. But that did not change the fact that Hakomoto, the mercenary insect dealer, had achieved something that the entomologists couldn't fathom.

'That's beside the point, however, for it is the moment of truth.' Max looked thoughtfully at the strange *Carabus* specimen. 'Did this collection never strike you as a little odd? All flower chafers, rhino and stag beetles, and then this *Carabus*?'

'I don't see what you're getting at.'

'My word, is it not obvious to you?' He chuckled, and angled the box so that its contents were now facing me. 'My question is – why no longhorn beetles? Or anything that develops inside *living* plants?'

I was utterly bewildered by this irrelevant gibberish, and it seemed as if the great entomologist had suddenly lost his mind. He was swivelling in his chair and, seeing me shaking my head at him incredulously, burst into a fit of laughter.

'Oh, good old Hakomoto! Mêdog, of all places!'

Far from being meaningless, these words triggered a moment of epiphany – Mêdog is a military zone! *How did he get in there?* The immaculate series of rare beetles, the incredible colour variations, the confusing anatomical features… all pointed to a truth so absurdly simple, yet so irrefutable.

Max's laughter echoed on in my head as my mind spiralled into darkness, awakening from what seemed like a short but terrible dream.

WARRIORS

A DANGEROUS ENCOUNTER

Hainan, June 2017

I shivered, and woke to an unusual silence. I was the only passenger left.

'This is the end of the road.' The driver turned around and grinned at me. 'I can't take you up any further, you have to get off here.'

I took up my net, and stumbled out of the minibus into a small clearing. The stifling heat of the air inside the bus had been too much to bear, but my sluggishness was immediately washed away by a cool breeze rising from the valley below. I had arrived at the jungle. Peering cautiously over the edge, I saw that the rocks crumbled away to form a sheer vertical drop of a hundred feet or more, tumbling down into the mighty torrents of the river beneath. The mist from the turbulence drifted up the steep forested hillsides, which thinned towards the bottom and gradually gave way to reeds and grasses clogging the roadside verges. But most spectacular of all was a mountaintop, or rather a series of peaks, looming over the foothills; strange formations of crimson clouds clung to their jagged forms, which rose like giant pyramids piercing the skyline. These great mountains stood as the highest peaks in Hainan, forming the centre of a vast plateau dominating the island's rugged interior. I stood in awe at this majestic sunset view, still baffled as to how I had travelled from a dusty town to this primeval landscape in the space of an hour.

The driver got out in his carefree manner and lit a cigarette. The sun was now sinking under the horizon, and the last rays of daylight dwindled as they slanted away from the jagged peaks, which were now becoming obscured by some ominous–looking clouds.

'It's going to rain soon. You can ask one of the farmers around here for accommodation, or camp out in the clearing.' He nodded at the DSLR camera hanging by my waist. 'I presume you want electricity though.'

My eyes were still fixated on the huge mountain, when I spotted a little ravine snaking its way down from the peaks. I scanned along its length, eventually following its course down to a rocky ditch crossing the clearing before me. All of this made me wonder if it was in fact a dried–up waterway.

'Where does that lead?' I enquired.

'Who knows? Could be one of the paths used by the local Hlai, could just be a dead-end hiking trail. I wouldn't put my money on it, son.'

'What are the Hlai like? They friendly?' I continued looking up at the strange path, wondering what kind of people could live in such a place. I had long known of the Hlai people, but only for some of their more unusual traditions – I had no idea what their reactions would be when they saw me.

'Humph!' he snorted, exhaling huge plumes of smoke from his little nose. 'They're the most honest people in the world; it's not them you need to look out for.' He gestured towards the peaks, and stared at me intently. 'Up there, people go missing all the time – poisonous snakes, vicious beasts, strange diseases – you name it. A couple of years ago a hiker found someone's shattered skull under a cliff, and it was already half-gnawed by porcupines. Couldn't even identify who it was.'

I smiled. I had heard such warnings all too often. 'Thank you, you have been of great assistance. I'll call you in three weeks' time.' I shook his hand firmly, and continued my way up the trail in search of lodgings.

A series of peaks looming over the foothills

I was in luck. About a mile up the dirt path the forest began to thin on both sides, and a small Hlai village of several dozen huts suddenly came into view. The locals were sitting on their porches finishing off dinner when I, with my huge butterfly net, must have appeared as some circus freak before them. They were uniformly tanned, small in stature and showed a typical emotional reticence towards outsiders. The village elders were more talkative, and told me that there was an empty hut further up the trail that guests sometimes stayed in. I thanked them, and soon came to a small thatched house tucked away behind a curtain of banana trees. The place looked clean and idyllic, and fitted the description of the guest house; I saw an old lady sweeping the porch with a broom, so I walked up to greet her.

I was met with a most inexplicable reaction. She initially looked up at me with a smile, which immediately contorted into a look of extreme shock as her gaze settled on the butterfly net. I did not think much of it at the time, as my sudden appearance must've come across as very strange, and I asked her about the availability of accommodation. After a few minutes she seemed to loosen up a little, and reluctantly agreed to let the place for 50 RMB a night.[1] Despite the mixed reception, I was more than happy with the arrangement: the inside of the hut was cosy, complete with a bed, electricity and everything a traveller could wish for. After spending a busy hour or so unpacking my equipment and clothes, I collapsed onto the bed for some much-needed sleep. A whole day of travelling had taken its toll, so that even the hard bamboo mat seemed like a luxury for my sore neck and back – but just as I

---

1   About £6 at the time of writing.

was about to close my eyes, a familiar voice began to echo in my head: *Lots of torching at night, especially on the damper evenings. And head for the peaks if you can!*

A sudden rush of adrenalin filled my veins, and the weariness I had felt earlier vanished instantly. I sprang out of bed and peeked out of the window. The villagers had retired for the night, and except for the odd cricket chirp there was an eerie silence. Beyond the village lay the dark forms of the great mountains, so vast and menacing that it was difficult to imagine the beautiful scenery I had witnessed only hours earlier. I began to remember why the night wilderness excited me so: the sheer uncertainty that permeated the senses, sharpening them while creating the wildest visions of fear and suspense. *I can't simply sleep this night away.* Every moment spent lying in bed would feel like another moment wasted, a missed opportunity to complete my mission – and tonight the conditions were perfect.

Before I knew it, I had a flashlight in my hand and was already losing sight of the village as I scrambled up the dirt path. Exploring the wilderness at night, or 'torching', as entomologists call it, is not unlike the masochistic act of watching a horror movie, except the heightened reality intensifies those emotions even more. You never know what is beyond the reach of your flashlight, or what creatures could be out there in the darkness – every crunch of the leaves, every quiver of the twigs, is enough to send your heart racing. You turn to see what is there, but it is often gone before your light hits the spot. It goes without saying that torching alone is fraught with hazards, but I rarely have the luxury of company; all I could do now was swallow my fears and bite the bullet, so to speak.

As I clambered over the barricades of giant buttress roots, I couldn't help but wonder how I'd managed to end up in a place like this. Only a week earlier I was having picnics with friends on Midsummer Common while watching cows grazing on the gentle English pastures, a far cry from a nocturnal venture into the treacherous jungle terrain. 'Probably still hung-over from May Balls!' I chuckled to myself. As a matter of fact, I much preferred being here, as I wouldn't—

A loud crash sounded from the path ahead. I froze. *It could've been a rotten branch, ripe fruit, palm frond… anything.* I tried to calm myself. *There's stuff falling from the canopy all the time.*

A steady rustling noise seemed to suggest otherwise, and it was intensifying as the source moved towards me. I could tell it came from a large and restless animal, as the footsteps were loud and hurried. *But what could it be?* I remembered seeing somewhere that Hainan was free from large animals due to its insular isolation, a theory I was very ready to believe at the time. Apprehension soon turned into curiosity as I realised this animal was no stealthy predator, so I decided to investigate further. When my flashlight fell on the path ahead, however, I gave a wild start as it illuminated a most hideous-looking creature.

Facing me was an enormous porcupine. Dazed by the powerful searchlight and quivering with agitation, he suddenly turned around and I found myself facing the hundreds of needle-sharp quills on his backside. Far from being cowardly, this was his final warning: should he be harassed any further, he would reverse into me at lightning speed and embellish my legs with dozens of broken quills. Those wounds would quickly become infected in the jungle heat, and that would be the end of my travels.

I switched off the light, and everything went pitch black. The porcupine rattled his quills. Inch by inch, I began to edge my way off the path, cringing as every leaf seemed to crunch with double its usual noise. I had no idea if some other nasty creature was coiled up in the bushes, but at that moment I couldn't think of anything other than to get out of the porcupine's way. I grabbed onto a large tree and began shuffling up its smooth trunk, desperately trying to get off the ground and hugging it for dear life. All those years of tree-climbing hadn't been for nothing, I guess!

A most hideous-looking creature

The wait felt like hours. Eventually, I heard the familiar rustling sound again and his waddling footsteps fading away into the distance. I had no idea how high up the tree I was, so I simply dropped off the trunk like a clumsy panda. A bush tripped me up as I fell, so that I found myself trapped inside what seemed like a botanical porcupine, scarcely more benign than the original.

'Off to steal crops, the little bugger!' I cursed. 'Bet the villagers won't be so kind to him!' I freed myself from the thorny bush and staggered to my feet. Blood was gushing out from the scratches on my forearms, which had become so tense from gripping the tree that I could barely close my fingers. Not a graceful start to my rainforest adventures, but a stark reminder of its inherent dangers.

Thankfully, I managed to pull myself together and continued the trek uphill, which became steeper and rockier by the minute. The nocturnal inhabitants of the jungle continued to test my nerves, fuelling my paranoia with strange illusions: my torch often caught the reflections of bright clusters of lights, which turned out to be the eight gleaming eyes of huge huntsman spiders as they patrolled the steep banks. Resting birds would suddenly crash out of trees shrieking with alarm, and I was never quite sure who was more terrified in such encounters. I was at my wits' end and about to call it a night, when something further up the trail caught my attention.

Why are those trees shimmering? I paused, and wondered what to make of this. A few feeble rays of moonlight sprinkled the path ahead, glowing like pearls in the darkness.

'Bizarre!' I muttered to myself. Light is a rare thing on the jungle floor even during the day, so seeing it at night was like seeing an apparition. I decided to go ahead and investigate. After finding the gap where the light penetrated the thickets, I fought my way through the dense vegetation and emerged unscathed on the other side.

I will never forget the thrill of discovering that place. A huge clearing unfolded before my eyes, so vast that it revealed the undulating forms of the mountain beneath; a row of trees stretched into the distance forming a neat fence around its perimeter, which suggested that this was the work of professional loggers rather than locals. There was a steady drizzle out here, which was completely unnoticeable earlier in the forest; the moon shimmered with a blue radiance, casting down the shadows of dark clouds as they passed over the crescent.

I scanned my searchlight across the ground, and instantly picked up the metallic reflections from several ground beetles. As Max had predicted, the muggy atmosphere triggered the activity of thousands of slugs and snails, which in turn lured out the nocturnal predators that preyed upon them. The sight of this swarm was a real confidence boost – *Carabus* was a predatory ground beetle, albeit of a giant kind, and there should be no reason why it wouldn't be here to capitalise on the banquet as well. In the space of an hour I collected at least ten different species of ground beetles. Most were about half an inch long and had shiny black shells, which occasionally glowed with a metallic lustre of greens and purples. One kind, however, was garishly dressed in a combination of black and yellow so that it stood out from all the rest. I had never seen such a beautifully coloured ground beetle before, and in the heat of the moment I neglected nature's most obvious warning sign as I reached my hand towards it – after all, it's not a wasp, so it can't sting. Right?

'Ouch!' I gave a yelp of surprise as my fingers burned with a scorching sensation. Leaving behind a high-pitched 'pop' and a plume of smoke from its rear end, the beetle had disappeared into the depths of a tussock nearby. A brown stain was left on my fingertips along with an awful lingering smell, but there was no stinger or mark to suggest that I had been stung. Alarm soon turned into elation, as I realised that I had just chanced upon my first bombardier beetle, along with one of nature's most remarkable events.

What had happened was a series of extremely complex chemical reactions: the beetle first secreted a fuel and an oxidant into a reservoir chamber, which formed an inert mixture for ease of storage.[2] When I picked it up, a small amount of this mixture was released into a sturdy reaction chamber, which was capable of tolerating high temperatures. The walls of this chamber were lined with enzymes, which triggered a violent reaction within the mixture to form a boiling solution close to 100 degrees Celsius. The massive pressure generated from the heat forced shut the valves leading into the reaction chamber, thus protecting the beetle's internal organs from the 'backfire', and exited through the rear as a controlled explosion. The tip of the abdomen could be rotated freely and the explosion aimed at any angle, since the shortened wing cases had left the tip exposed. As if one explosion wasn't impressive enough, this cycle was repeated at an astonishing rate of 500 pulses per second – yes, that squeaky 'pop' was in fact the sound of over seventy individual explosions! Even with modern technology, the firing speed of our weapons truly pales in comparison to this biological machine gun.

Regrettably, I did not see another bombardier beetle that night, and the giant ground beetle *Carabus* was nowhere to be seen. My epic nocturnal adventure was brought to a sudden halt by a flash of lightning that tore through the night sky, which left me running for my life down the mountain. The thick canopy helped to shield much of the raging storm above, so that I was able to get back to the village relatively dry and collapse onto my bed, at long last.

---

2   So complex, in fact, that creationists often use the bombardier beetle as 'evidence' for intelligent design, arguing that such a complex mechanism could not have arisen through evolution. One can write an entire book about the intricacies of the mechanism, so I shan't delve deeper here; a more complete explanation is offered in the wonderful book *For Love of Insects* by Thomas Eisner, which I would thoroughly recommend to anyone with an interest in invertebrates. It is an especially fascinating read for the more chemically minded.

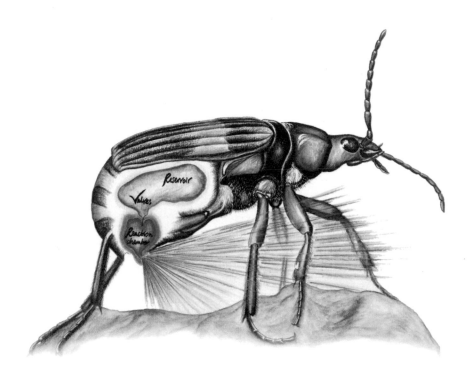

One of nature's most remarkable events

THE DELUGE CONTINUES

Much to my sorrow and frustration, the thunderstorm lasted more than just a few hours. For another three days and three nights rain poured down from the heavens, and the forest dirt paths were transformed into murky torrents that raced down the sides of the great mountains. Every time, just as the downpour seemed to have subsided, another thunderbolt came crashing down to warn of another storm approaching. Deep down I knew that this rain was the jungle's lifeblood, but at the same time I was disheartened that it was keeping me at bay. There was nothing I could do until the storm passed but wait, and plan my expedition to the peaks.

I took this time to familiarise myself with the customs and traditions of the villagers, which turned out to be remarkably familiar: they had their own Hlai language which I found completely unintelligible, but I was pleasantly surprised to find that they spoke perfectly good Mandarin, even the village elders who had probably never left that mountain their entire lives. The Hlai lifestyle revolved around farming rice down in the paddy fields near the river, so they spent relatively little time in the forests except when they needed firewood and medicines; nonetheless they had a good knowledge of the local flora and how they could be used, which I was eager to learn as botany was never my forte. They were an industrious lot, and the poor weather frustrated them as much as it did me since it meant much of the farming work had to be put on hold. As the villagers got to know me better, they became curious about the purpose of my trip and agreed to help me with my search.

On the morning of my fifth day there, the storm finally calmed and the clouds dissolved to reveal a most brilliant blue. I did not make straight for the peaks, however, as an army of mountain leeches would await me there, roused by the recent storm. The villagers promptly took up their tools and headed down into the paddy fields, along with the merry children who perched atop the water buffaloes; they invited me to go with them, as they told me there was a 'hard-shelled bug' peculiar to the plantations which would interest me greatly.[3] I gladly agreed, and together

---

3   At the time I could only guess that they were referring to a beetle, but of course there are other insects that are heavily armoured, such as the true bugs. As an aside, the English word 'bug' is an over-abused word that is generally used to encompass all arthropods; strictly speaking, true bugs only refer to the order Hemiptera, which contains aphids, cicadas and some other insects with sucking mouthparts.

we worked our way down the mountain towards the lowlands. The landscape here was more open and showed significant signs of human disturbance, with the rugged hills giving away to ponds, coconut trees and sugarcane plantations. Nevertheless, there were some pleasant surprises that I encountered along the way, a sign that the jungle had not lost its grip on these territories.

One of the most remarkable plants that grew beside the paths was the giant taro, which thrived among the weeds and bushes. Their enormous foliage looked especially attractive after the big cleanse, and many of them were also sporting white flowers and bright red fruits. This overall exotic appearance has made them very popular as ornamental plants, but they can be very dangerous if mistreated – every part of the plant is ridden with raphides, microscopic crystals of toxic calcium oxalate that can tear open the digestive system of any herbivore brave enough to eat it. Like its smaller cousins, the giant taro produces tubers that form the staple of many tropical diets; prolonged cooking is required, however, to destroy the raphides within and render them palatable. The villagers never ate them due to their violent properties, and the plants were left to grow into colossal forms that towered above the gloomy undergrowth.

The walk down to the river was much trickier than expected, as the descent from the road was steep and made extremely slippery by the recent deluge. We cautiously edged our way down the slope, clinging onto tussocks of grass to maintain balance and footing. Even the buffaloes managed to make it down in one piece, and showed a nimbleness which I hadn't given them credit for. When we reached the river, the team split into two: the men were to harvest the last of the spring batch from the paddy fields, and the children were to remain near the waterhole and look after the buffaloes. I naturally went with the

The giant taro

buffaloes as I was curious to see what they did in their spare time – replanting for the autumn batch was about to begin, but all the ploughing had already been done, so the animals were enjoying some well-earned rest. The waterhole was overflowing into the adjacent meadows, and created a muddy heaven for the wallowing buffaloes and playful children. The reek of pond mud and swarms of mosquitoes was rather more than I could bear, so after receiving multiple bites on my neck I decided to head for the plantations and find the 'hard-shelled bug' I had heard so much about.

The first telltale clues were neat, circular holes in the ground almost two inches across, which I speculated to be the exits through which the adult insects emerged. They were made, no doubt, after the torrential rains had passed, and my pace quickened at the thought of some giant beetle lurking in the plantations nearby. It was not long before I found the culprit sprinting across the dirt path ahead, a diabolical creature unlike anything I had seen before. The three-inch-long monstrosity had the curved jaws of a stag beetle, the twig-like antennae of a longhorn beetle and the thickset body of a rhino beetle, all merged into a jet-black tank covered with spines and claws. As soon as I seized it, it began to slash at the air with its jaws, desperately trying to slice my fingers open while hissing with a sinister sound.

Holding the menacing creature between my fingers, it was hard to imagine how something so primeval-looking could have emerged from a place so artificial. I looked up at the muddy track, and noticed that there were several more of these things crawling and flying in the distance. I suddenly remembered that the villagers had told me they never saw such insects in the forests, and that they were becoming more common in the plantations year on year. I suspected that these beetles had probably begun their lives by feeding off the roots of sugarcane, living underground as translucent grubs for much of their lives just as many other beetles do, before emerging from the holes I saw earlier. Further inspection of the antennae confirmed that this was a longhorn beetle, despite its rather unusual appearance and life cycle.[4] The reason for their absence from the jungle was probably because such swathes of host plants simply didn't exist in nature – the rows upon rows of sugarcane now served as the perfect breeding ground for a population explosion. With few predators daring to tackle such ferocious prey, these beetles had now become serious pests of plantations in the lowlands, destroying acres of healthy plants every year.

---

4   Longhorn beetles are usually slender in appearance and have long antennae, which give them their name. The larvae of most longhorn beetles develop under bark or inside wood, often within the trunks of living trees.

A population explosion

I decided it was finally time to do something socially beneficial with my hobby – after all, removing these vermin would save the farmers a lot of pesticides and grief. I ran up and down the path catching these vicious insects, and put them in separate containers to prevent them from biting each other: though not cannibalistic, these beetles wouldn't have a moment's hesitation in severing their siblings' legs or antennae. When I returned to the waterhole the children were already bored of playing on the buffaloes' backs, and my bounty of mini-beasts presented a welcome change from their everyday games. Initially, I was worried to let the children play with them, as they could do some serious damage, but they soon learned to handle the critters with all the dexterity and caution of a trained entomologist. I amused myself by feeding various plants into the beetles' jaws and watching them snip the stems effortlessly like secateurs, while the children tethered them with strings so that they flew around like toy helicopters. The options were endless, and the beetles provided continuous entertainment to satisfy our childish playfulness.

During the course of the afternoon, I noticed that one boy in particular, Jin, behaved very differently from the others. He stood over a head taller than the rest of his peers, and was of a much calmer disposition; I never saw him laugh or scream like the others, but the fierce sparkle in his eyes spoke of an exceptional liveliness and intelligence. While the other children swam around throwing mud at each other, he would sit by the pond flicking through his tattered copy of *New Concept English*, or looking thoughtfully at the clouds shape-shifting across the sky. The boy took an immediate liking to me, and was very eager to learn more about the details of my mission. He must've only been around thirteen but had amassed an incredible amount of knowledge on his own – as a matter of fact, he pretty much succeeded in communicating with me in English, which made me wonder what he would've been capable of if he had had a formal education. When I showed him the beetles, he surprised me by taking out one which he had collected earlier and kept in a plastic bottle. It had shorter mandibles and a distended abdomen, and I only realised later that it was the much rarer female of the same species.[5]

It was already dusk when the farmers finished their work in the paddy fields, and as the team reassembled we began to make our way back to the village. The men were now carrying hefty

5   The mandibles of an insect are its jaws, which are usually used to chew food. In some male beetles the mandibles are so enlarged that they can no longer be used in feeding and simply serve as weapons for combat, the most famous example of this being the stag beetle.

baskets of harvest with poles resting on their shoulders, which bent under the weight of the burden; hungry and exhausted from an intense day of work, the villagers were eager to return to a hot meal awaiting them back home. Soon we arrived back at the small clearing and the start of the village path, which marked the jungle's frontier – leaves, branches and countless other kinds of debris were strewn over the ground, washed down from the jungle by the recent deluge. Such flood debris is often full of interesting things, and gives a fascinating insight into what goes on in the gloomy jungle undergrowth. As I scanned the ground to make sense of the mess, I noticed a peculiar object poking out of the ground right by my feet.

'Is that a twig?' I murmured, leaning over to take a closer look. What drew my attention was a reflection, which I thought came from a piece of metal. I took the dark object from the mud and swivelled it between my fingers, trying to see where the reflections came from. As I tilted it against the dwindling light, flashes of rainbow colours seemed to appear from nowhere. The truth suddenly dawned on me and, for a moment, I could have sworn that my heart stopped beating.

It was an insect's leg, but monstrous – outstretched, it measured nearly five inches long, and was equipped with spikes and claws that seemed far too large for any insect. At that moment I felt like the fossil-hunter who had just discovered the arm of *Deinocheirus*, as I was equally baffled by my find.[6] I trembled at the thought of discovering its owner, and looked up at the dark

6   For almost fifty years, the only fossils of *Deinocheirus* were a pair of eight-foot-long arms. These sparked huge interest and debate among the palaeontology community, with some believing it to be a predatory dinosaur that used its huge claws to tear prey apart. In 2014, the dinosaur's appearance was finally resolved after the discovery of a skull and two headless skeletons.

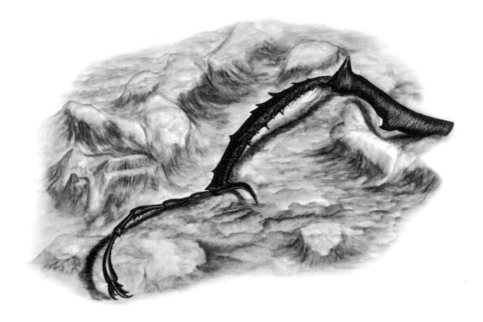

A peculiar object

shapes of the foothills towering above me – somewhere in there, in the depths of the jungle, lurked a creature that might have never been glimpsed by anyone before me. I carefully folded the limb into a glass vial, and followed the villagers as they disappeared into the forest.

MEMORIES OF THE PAST

I barely slept that whole night. The limb I had discovered was as troubling as it was exciting, and I tossed and turned while trying to imagine what on earth it could be. I sat up in my bed and gazed up at the starry sky; in my mind there was a feeling, an idea that was stirring from the hazy depths of my memories – yes, I had seen something like this before, many years ago. It was in a peaceful land of temples and forests, where I became acquainted with some of the greatest warriors of the animal kingdom.

Inspired by tales of the great adventurers of old, I had become infatuated with the idea of exploring the unknown – in my mind, there was nothing more thrilling than the prospect of becoming a pioneer of the wilderness and discovering a species new to science. My family often tried their best to satisfy my unrealistic desires of thrill-seeking, but there is only so much freedom you can grant to a reckless teenager! As a result, most of those initial escapades were mere ambles within the numerous country parks that dotted my hometown in South China. Vast forests of bamboo thrived in the lush sub-tropical climate, and the cool shaded paths that ran along the hillsides offered a peaceful retreat from the scorching summer heat. Even now, I often think about the extraordinary things I found on those day trips, many of which I never saw again in my later travels.

A striking feature of these temperate woodlands was the abundance of sap-bleeding oak trees, which any entomologist would hail as a godsend. The sickly frothing liquid that oozed from the trunks was often a sign of bacterial infection or of some pest within the tree, and that the mighty organism was nearing the end of its days. Insects were only too happy to take advantage of this deterioration, and a whole host of them could often be found feasting on these 'sap-runs' – dazzling flower chafers, swift Nymphalid butterflies and ferocious giant hornets. This 'aerial nightclub' inevitably brought much drama, as the clubbers squabbled incessantly over the females and the best drinking spots. Amid all this decadence the rhinoceros beetle thrived, using his bulk and brutish strength to bully the other insects into submission. The largest males were, naturally, the best at fighting, and guarded their sap-runs with the utmost confidence – I often saw them with their heads buried deep within the fissures of the trunks, oblivious to the giant hornets that constantly harassed them. Stings, claws and jaws were simply useless against the beetle's armour.

Pound for pound, the rhinoceros beetle is perhaps the strongest animal on the planet. They have a rather curious way of fighting, where the sparring male attempts to 'scoop up' his opponent with a long horn growing on his head which branches off at the end like a garden fork. Despite its menacing appearance, this implement is nothing more than a glorified crowbar – once underneath its opponent, the beetle rears itself up with powerful front legs and raises its head with such force and swiftness that the opponent is often sent flying into the air. Imagine that, being able to fling someone into the air just by craning your neck back! To counter such attacks from opponents, these creatures have developed a most unbelievable grip on rough surfaces; it is only when you try to catch one that you begin to appreciate their immense power, as it is surprisingly difficult to even lift them off the tree trunk. Once you have one in your hand, it is not to be handled roughly – it is almost impossible to hold one in your grasp, as they are covered with a multitude of spikes and claws that can easily rip the skin. This impressive arsenal, paired with a fierce temperament, accounts for their popularity in Japan as pets, where a bout between two equally matched males makes for the perfect gambling game. More experienced keepers can even encourage the beetles to breed, producing many generations of offspring with increasingly favourable characteristics.[7]

---

7 Nowhere else in the world is 'insect culture' as popular as in Japan, where many children spend their holidays catching beetles and keeping them as pets. Such experiences inspired the game designer Satoshi Tajiri to create the globally successful Pokémon franchise.

One trap that did work on occasion

After seeing the incredible effectiveness of tree sap as an attractant, it later occurred to me that I could easily fabricate some kind of trap that exploited the insects' craving for sugar. At first, I would often go to great lengths in designing these, using various inverted funnels and flaps to ensure that the inmates didn't escape. However, as is often the case in life, intricacy doesn't necessarily correspond to efficiency, and many of those contraptions that I deemed so ingenious at the time now seem rather laughable and contrived. There was one trap which did work on occasion, and it turned out to be the simplest of them all: a rotting banana tied inside an old sock, which was then hung off a branch. Using this method, I was often able to catch stag beetles, which are rarer than their rhinoceros cousins but equally aggressive. If anything, they are capable of far more damage as their mandibles are razor sharp and can clamp together with the force of a crab's pincers. Indeed, on more than one occasion my finger found itself trapped in the beetle's vice-like grip, and apart from gasping for air and waiting until the little devil had had enough fun, there really wasn't much else I could do. Trying to prise the jaws open (with just one hand!) was futile, as the beetle would only become more irritated and tighten its hold.

The warmongering lifestyle of these beetles, though flamboyant, came at a heavy price. Many of them were killed in combat, and towards the end of the season in August it was difficult to find one that had not been maimed. When the autumn wind began to blow in September, the forest floor had become littered with the glistening shells of beetles, a bygone memory of these warriors' brief lives.

JESTERS

THE PINNACLES UNVEIL

I lit my bedside gas lamp, and took the leg out of its glass vial. The harsh yellow light intensified as if reflected off the strange object, which seemed more like a bronze artefact than an organic limb. The tibia was exceedingly spiny, armed with countless sharp points as if it were a medieval instrument of torture.[8] At first sight the claws looked similar to those of a rhinoceros beetle, or something related to it, but upon closer inspection they possessed multiple hooks, which was a feature I had never seen before. Besides, if this mysterious insect was of those stocky proportions, it would have had to be over seven inches long to own such limbs – a monstrous size that is beyond that of even the biggest

---

8   An insect's tibia is the equivalent of our forearm, or crus.

scarabs in the world.[9] Such a discovery would surely shock the entire scientific community.

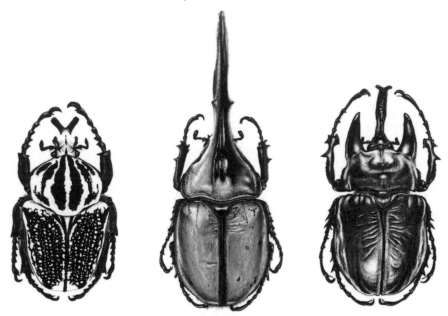

Beyond that of even the biggest scarabs in the world

When the roosters began to crow at the break of dawn, I decided to give up on sleep and make an early start to my expedition. A blue glow was rising from the horizon, and the scattered wisps of dark clouds were signs of a pleasant day ahead. I packed my bag with filming equipment, and sat down to enjoy my last can of sweetcorn. Water wasn't a problem as I was free to use the village well, but my rations were fast running out – the most difficult part of the expedition was about to begin, and I needed to find a source of food. The locals ate almost exclusively

---

9   In the illustration on this page are three of the largest scarabs in the world, from left to right: *Goliathus orientalis*, *Dynastes hercules* and *Megasoma actaeon*. The beetles are represented at roughly half life-size.

rice, which was cooked over a fire inside sections of bamboo and then served with a small side dish; these were often boiled concoctions of vegetables, tubers and game of questionable origins, which I was less than enthusiastic to try. Thankfully my host, who turned out to be Jin's grandmother, pointed me to an orchard at the edge of the village, where I could pick whatever I pleased and pay her afterwards – there seemed to have been some misunderstanding between us at the start, as she turned out to be as friendly and as helpful as any other villager. I provisioned myself with a bunch of ripe bananas, fresh from the tree, and made my way down to the river crossing.[10]

The river crossing, as I called it, was nothing more than a giant concrete slab sprawled across the river. It signified the boundary between the mountains and the lowlands, where the jungle fought desperately against its shrinking perimeter. The landscape here was scarred with old quarry pits, ramshackle huts and abandoned townhouses, with nothing to shield the tropical sun except the few shrivelled saplings that dangled from the rooftops – it was an ugly place, where the weeds grew waist-high and swallowed up the desolation that human progress had left in its wake. My lips contorted into a grimace, and my pace quickened as I attempted to leave the dismal scene behind me.

Out of the corner of my eye, I noticed that something was awry. A plant was growing from a pile of rubble, and as I gave it a sideways glance I noticed that an object was making one of the leaves droop ever so slightly. Be it a gift or curse, naturalists are probably the most easily distracted people on this planet; we often

---

10 The banana 'tree' is actually a giant herb, made up of the bases of the huge leaf stalks. One 'tree' can only ever produce one bunch, and so is usually cut down after the bananas are harvested.

see trivial details that others overlook or ignore, and take great pleasure in examining them. After years of 'looking' in this way, many develop an apparently superhuman ability to spot critters, no matter how small or well camouflaged. It might just be a tiny shadow cast on a leaf, or a twig that's set at a somewhat unnatural angle... These details jump out at us and, for a moment, they become the focus of all our attention. Cautiously, I took the leaf by its stem and flipped it over.

Clinging onto the underside of the leaf was the most perplexing object I had ever seen: two sharp horns protruded from a translucent blob that seemed neither to move nor react to the disturbance. Excitement and confusion flooded my senses, and my heart began beating violently as I reached my hand towards it in apprehension. It was an animal, of that much I was certain; yet I could not find any eyes, legs or antennae that would clarify its identity. For all I knew, it could have been some marine creature lifted out from the depths of the ocean, as it would certainly have seemed less out of place there. I held the blob by its horns, which were surprisingly rigid, and gave it a little tug. It was firmly stuck to the leaf, and didn't budge an inch. All I could see was a gentle pulsation of blues and greens through its translucent shell, the only sign of life in something which could have passed as a plastic toy.

Marvelling at the creature alone was not going to be informative, however, so my next thought was about how to bring it back for more detailed observations. I rummaged through my bag looking for the box of glass vials, only to realise that I had left it out in the morning when I was examining the beetle's limb. That box of vials was the one piece of vital equipment I needed to bring, but I had forgotten it, in true naturalist style. I needed to turn back, but how to carry this thing? The globule remained motionless on the leaf, so eventually I came up with the idea of inflating a plastic bag and placing the entire leaf inside, tying up the opening with a knot. Fearing that it might become crushed inside my rucksack, I secured it to the ribbon on the top of my straw hat – with this 'airbag' perched on the top of my head I ran back to the village to fetch my collecting gear. It would seem that improvisation is a naturalist's forte, developed through years of dealing with our own forgetfulness! Thankfully none of the villagers saw me in this comical state, or else I would have been their laughing stock for many years to come.

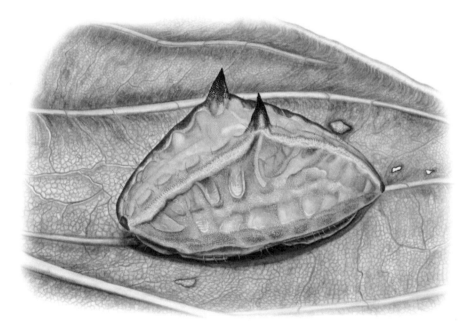

A translucent blob

Many of my findings in the jungle completely baffled me, so much so that at the time I could form no theory as to what they were. After careful observation and inference, however, they usually turned out to be related to something I was familiar with, albeit of exaggerated proportions. I found the whole experience endlessly refreshing, as if I were a child venturing into the wide world for the first time. Every nook, every cranny had hidden gems waiting to be discovered, and when the revelation came – ah, it was a feeling beyond my wildest dreams! The curious blob made me wonder how much was still left undiscovered in this jungle, and what lengths I would have to go to in order to find these relics. For me, the answer was an obvious one: if *Carabus* was anywhere, it would be up there on those cold misty peaks, far removed from the stagnant heat of the lowlands. The question now was how to get up there, a problem that turned out to be far more difficult and dangerous than I could have ever imagined.

My delay on the road, though unexpected, turned out to be most convenient. The first rays of sunlight had crept above the skyline, and I was greeted with an impressive view: the pinnacles of the great mountains, which had been shrouded in mist ever since I arrived, were now in plain sight – five jagged peaks cloaked in darkness, reaching like a great hand into the pale morning sky. About halfway up the western slope, the steep river valley was cut off abruptly by a sheer cliff face, producing a spectacular waterfall that hung as a white ribbon in the sea of greenery. It was here that I would be heading, and encountering more creatures of the queerest and most ludicrous appearances.

JOURNEY UPRIVER

The road by which I arrived at the jungle was a rather crude one, as it turned into a dirt path abruptly at the river crossing – there was no indication of where it led or how much longer it would go on for; indeed, the trail seemed to wind away from the mountains and be headed for some foothills further south. Even worse, it could have just been a dead end after a few miles. Therefore I decided against following the road and took the only path I deemed reliable – the jungle river, the source of which must be somewhere near the summit. Looking at the sheer size and volume of the waterfall further upstream, I assumed that the flow of the river would remain unbroken throughout, carving a natural path through the dense forest for me to follow. Such routes are often the best, and only viable, way of navigating through rainforests.

Fortunately for me, the aftermath of the recent deluge had already subsided, and the water level was now more or less normal. The river channel was full of dead branches and palm fronds brought down by the recent typhoon, which crashed between the boulders as they were carried downstream by the raging torrents; I was grateful that I didn't have to negotiate the waterway itself, as the banks of the lower sections were still lined with a gentle gravel bed. As the monsoon season progressed, however, the river would gradually swell up to flood these areas, even spilling over into the jungle. Then, taking this route would be as good as suicide.

I was lucky enough to have this gravel path for a significant part of my journey, and all along the way I was greeted with the most unbelievable scenery: the airy space above the river had become filled with the bustling flight of dragonflies, butterflies and swifts, which put on a display of aerial acrobatics that grew ever more elaborate as the rising sun warmed the atmosphere around them; the canopy above the banks, which curved inwards to make the most of the precious light, supported a myriad of vines and creepers that hung like botanical curtains from the lofty branches; giant palms and tree ferns dominated the understorey, whose parasols of fractalesque foliage formed a strange contrast with the tangle of vines above. Such organisation and complexity was reminiscent of some futuristic highway – yet all of this had emanated from the ground far before humanity ever existed, and continued to function perfectly without our interference.

A huge strangler fig

Beyond the bridge, about two hours into my journey, the vegetation suddenly thickened and the gravel path came to a halt completely. I took this as an opportunity to have a water break, and to observe the commotion in the branches above: I was under a huge strangler fig, whose towering form overhung the river and cast a pleasant shade upon the waterfront. On closer inspection, I discovered that the entire weight of the crown was supported not by a conventional trunk, but by an intricate meshwork of roots that reached down from the great heights where the organism must have originated. Like the writhing tentacles of some grotesque sea monster, these aerial roots had been used to envelop and choke its victim – no trace of which now remained. The central hollow tube, where the unfortunate host tree once resided, stood as a stark reminder of the brutality involved in this slow and steady war, and as a testament to the unforgiving forces of time. Yet from all this death and decay there sprung forth a most vivacious community – high up in the canopy, the ripening figs had attracted animals from all across the jungle, who now congregated here to make the most of the daily offering. It was an excellent opportunity to see some of the more reclusive inhabitants of the jungle, and to take some photographs which would aid me in my future artwork.

Through the telephoto lens I saw, perched on the tips of the branches, numerous colourful birds whose names I did not know; the plumage was predominantly a leafy green, but the head was decorated with a striking combination of reds, yellows and indigoes. Giant hornets, no smaller than hummingbirds, were buzzing around the rotten figs and lapping up the fermenting cocktail oozing from the fruits. Further down the tree there were several rhesus macaques, likely members of a larger troop nearby, which were typically noisy and squabbling over the bounty. Amid the chaos I glimpsed a dark shape darting up the tree trunk in a peculiar jerky fashion, and with such swiftness that I couldn't even get it in focus. My first impression was that it was a Mustelid – but it was pressed flat as a pancake against the tree, so that it had lost all the gracility typical of such animals.[11] I circled the tree in an attempt to get a better view of the creature, but

---

11 Mustelidae (or Mustelid) is the scientific name for the weasel family, which also contains martens, badgers and otters.

it was simply impossible to see its squashed profile against the bright sunlight. Eventually I had to back off twenty yards or so to get a better view, and as the creature emerged onto one of the boughs it finally reassumed its natural form. I took a few rapid shots at an extremely high shutter speed, and leaned back from the viewfinder to look at the images I had captured.

I could scarcely believe what I was seeing. Instead of some marten, the animal running around on the tree was clearly a squirrel, but of the most gigantic proportions – from nose to tail it must have been over three feet in length, as it dwarfed all the birds that it managed to scare off. Even the macaques kept their distance, presumably out of fear of the giant rodent, and began shrieking with alarm. The squirrel paid no heed to the monkeys' snarls and simply jolted ahead, flicking its huge bushy tail up and down restlessly; apart from the obvious size disparity, its bearing was remarkably similar to its grey cousins from England.[12] In a blink of an eye, its nimble footwork had already carried it up to the tips of the branches, where it began to gorge itself on the fruits that the macaques couldn't reach. I stood there speechless, taking shot after shot of the strange beast and marvelling at its agility as it leapt between the branches.

A fifteen-minute water break turned into a two-hour diversion, as they often do on these trips of mine. The fig tree provided an endless source of new spectacles as I absorbed myself in watching the comings and goings of the canopy inhabitants, most of which were new to me. To add to my distraction there was the river,

---

12   Sadly, the grey squirrel which looks so adorable with its fluffy coat and round ears is far from benign. Since its introduction to the UK in the late nineteenth century, it has wreaked havoc over the island, displacing the indigenous red squirrel over much of its range through disease and competition over resources.

A squirrel, but of the most gigantic proportions

which placed me right in the middle of a flight path. I would turn my head whenever I heard something buzzing past me to examine its flight pattern, and net anything which seemed remotely like a beetle – given the sheer number of insects in the rainforest, I think it was fair to say that I was constantly distracted. Most of the suspects turned out to be bulky, metallic blue carpenter bees, which flew with a heavy buzz reminiscent of flower chafers. The most convincing impostor of them all, however, was a plump bug of the most astonishing appearance – a part of its thorax had been enlarged to such an extent that it formed a carapace covering the entirety of its abdomen.[13] If it weren't for the lack of a line down the middle, one could have easily mistaken this modification for a pair of hardened wings and assumed it to be a beetle.

13   The word 'bug' is used here in the strict sense, as this insect does indeed belong to the order Hemiptera.

I was pleased to find that these bugs were abundant among the camellia bushes, which thrived in the dim, humid environment of the riverside undergrowth. Among the mass of dark green foliage, the reds, blues and whites of these brightly coloured clowns glowed like neon lights, making them extremely conspicuous – indeed, those colours reached such extraordinary levels of saturation that I thought they were fake at first, and it was only after catching a few that I convinced myself of their authenticity and identity. Here was an example of aposematic warning, where the brilliant colours were used to advertise the animal's distastefulness – the bugs left a most horrendous smell on my fingertips, which was reminiscent of the sort produced by stink bugs in the temperate parts of the world. Without this cocktail of noxious aldehydes, these morsels probably would've been picked off by birds long ago.

Extraordinary levels of saturation

As much as I enjoyed my diversion, there were more pressing matters at hand. Dark clouds were now beginning to form around the peaks, and it was only a matter of time before they would materialise into a storm. If the rainfall up there turned out to be severe, deadly flash floods would occur without warning further downstream, where I found myself standing. I had lost far too much time already. The falls couldn't have been much more than a couple of miles away, as I could even make out the individual cascades with my binoculars, but beyond that? I hadn't the slightest clue. The terrain ahead rose sharply into the jungle – into a land of tangled bushes, sheer cliffs and treacherous ravines, continuing in this way for as far as the eye could see. A few rocky outcrops jutted out towards the pinnacles, splashes of greys and browns which served as the only breaks from the monotonous greenery. Just as I was lamenting why the mountains here had to grow so big, a most extraordinary noise resounded through the jungle.

I froze, completely shocked by its sheer loudness and strangeness. There was an unnerving, almost primeval quality about the sound, which I can only describe as a low-pitched siren that trailed off into a long note. After a moment's pause the bombastic melody started again, repeating each note with a mechanical precision and regularity – it was difficult to pinpoint the source as the volume was so overwhelming, but as I rushed towards a clump of bushes the song ended abruptly in a rasping wail, and all went silent. I rattled the bush with my net handle, but nothing ran out. I turned in all directions, half-expecting a *Gastornis* to crash out of the undergrowth, but there was nothing.[14]

14 *Gastornis* was a huge flightless bird that lived in the rainforests of the Eocene, around 50 million years ago. It stood taller than a person and likely preyed upon small mammals, giving rise to its nickname 'the terror bird'.

Whatever was making that noise seemed to have vanished into thin air, without leaving the slightest trace.

Then, as if by telepathy, a hoarse hum began to resonate from all directions – the melodies followed, gradually combining and swelling into a deafening chorus that filled the jungle atmosphere. It seemed as if the mountain had suddenly become a gigantic organism of its own, singing through its caverns and telling its tale through this mysterious and ancient song. In a hypnotic trance my legs carried me forwards through the currents, the sirens drawing me ever deeper into the heartland of the jungle.

BEYOND THE WATERFALL

The disappearance of the gravel path forced me to veer into the middle of the river, which was becoming narrower and steeper by the minute. The closer proximity between the boulders meant that the waterway was at least negotiable, and the flow here was a mere trickle compared to the mighty torrents downstream. I found myself hopping and clambering over the boulders, sometimes even climbing up large slabs where the river cascaded off a ledge. It was a dangerous business, as the bloom of algae had just been moistened by the recent flood and formed a slimy gloop coating the dark rocks – I slipped into the river many times, almost twisted my ankle on several occasions and became drenched from the waist down. There were many times of despair, when I was confronted with a crag so steep that I was convinced this would be the end of my journey, but I somehow managed to press on, either by diving back into the forested banks to pull my way up on vines or by using the saplings growing from the fissures as handholds. There were times when I was hanging off just one branch while tiptoeing up the rock face, so should the foundations of the plant have yielded then the consequences would have been unthinkable. But now was not the time for fear. I had come so far, and every step I took towards the summit meant another step closer to my quarry – *Carabus*, the assassin haunting the pinnacles above. I knew it was up there somewhere, and if Hakomoto could catch it, then I could as well. Fuelled by a relentless rage, I climbed ever higher up the ravine and towards the falls.

It is perhaps worth mentioning that I heard the strange noise on several other occasions, sometimes emanating from deep within the jungle, but every time I closed in on the source the sirens ended abruptly with that same rasping wail, with absolutely no movement in the vegetation or any noises to suggest that the culprit had escaped. For all I knew it was sitting right next to me, and either had such brilliant camouflage or nerve that it didn't even bother to escape. Other than that, however, the remainder of the trek turned out to be rather painful and uneventful. Throughout my clamber in the denser parts of the jungle, I did not catch anything worthy of mention – I found this to be true later in my travels as well, in that a denser forest didn't necessarily mean a more productive one. Sometimes it seemed as if the thick vegetation was suffocating the whole place, clogging up the space between the trees and not giving the inhabitants enough room to breathe. There was so little light and air under these thickets that it appeared completely devoid of flying insects – save a few tiny fungus gnats, whose maggots gnawed away at the plentiful supply of mushrooms. During these quiet periods of calm the scarcity of insects left me wondering if a plane had just passed over and sprayed the whole mountain with pesticide. Then, in an instant, a plethora of strange creatures would appear seemingly out of nowhere, leaving me clutching several beetles in each hand and wishing I had more to catch them all. Amid the chaos I often felt overwhelmed, frantically chasing after the flying ones with the net in one hand while fumbling to bottle up the recent catches with the other. Such was the situation I found myself in when I finally reached the end of the ravine and was faced with a colossal tiered precipice several hundred feet high.

Resembled a deep-sea creature

I had arrived at the waterfall. Powered by the recent torrential rains, countless cascades hurled themselves off the sheer cliff face and came crashing down with a deafening thunder, drowning out the melodious birdsong of the surrounding forests. An open space surrounded the base of the falls, overgrown with water convolvuli and other semiaquatic plants which flourished in the waterlogged edges of the plunge pool. Among the herbs and tussocks, I was confounded by yet another enigma, which resembled a deep-sea creature even more than the strange blob I had found earlier – a spiky, trilobite-like body with an intricate black tail, which reminded me of some contemporary bourgeois sculpture. I must admit that I was rather reckless to touch it, as many caterpillars can sting with their spikes, which could have left me in pain for days. Fortunately, the little thing turned out to be totally harmless, and under my persistent pestering did nothing more than flick up its tail and cover its body with the feathery mass as if it were a shield. What alarmed me was when I accidentally poked the tail and a piece of it fell off, but the animal didn't show any signs of pain or discomfort. As the piece of 'tail' crumbled between my fingertips, I burst out laughing as I suddenly realised what it was.

This 'tail' was not a part of the living organism, and from the consistency I guessed that it was probably the creature's dried faeces. So impressive was the scale and rigidity of the structure that I found it difficult to believe how the larva had made it within its own lifetime. I looked up at the convolvulus patch, and noticed that there were several others munching away at the leaves, all at various stages of development. These were, in all likelihood, the larvae of some beetle, as a few had already pupated and looked very much like the pupae of a ladybird, only several times larger. A shield of faeces may have been enough to keep birds and lizards at bay, but I wasn't fazed in the slightest – I was far too curious about the beetle's identity to worry about trivial things like hygiene. One after the other, I picked them up and put them into the glass vials, making sure that there was enough food in each one for them to complete their development – I only collected a few, as I did not want to eradicate this beetle from this environment. After scouring the whole patch and bottling up about a dozen larvae and pupae, I harvested some of the plants themselves and wrapped them up in a plastic bag, so that if the larvae's appetites turned out to be particularly voracious then at least I would have some reserves. Indeed, water convolvuli are also known as 'water spinach' and are a delicious vegetable that I could have eaten myself. As I waded through the waist-high vegetation, harvesting bunch after bunch, I saw a flash of light rising from one of the leaves ahead.

He took off clumsily like a mini-helicopter

If I could have seen myself in the mirror, I would have probably seen my jaw on the floor. A nugget of gold – no, not yellow – pure, blazing gold, suddenly spread its wings and took off into the air. For a moment I was dumbstruck, but as my senses returned I swung the net at it wildly – unfortunately, it was too late. The flying piece of gold drifted lazily across the pool, and vanished into the forest on the opposite side. I could do nothing but mourn as the object disappeared right before my eyes, but just as I thought the situation couldn't get more hectic, a flash of colour made me turn abruptly to face the waterfall. I immediately spotted a large longhorn beetle crawling up one of the rocks near the bottom – I pounced out of the swamp like a tiger after its prey, and not a moment too soon. The wing cases opened, and he took off clumsily like a mini-helicopter, buzzing loudly as he crashed around the vegetation in a futile attempt to rise up the cliff face. With a sweep of the net and a turn of the wrists, the vicious beast became hopelessly trapped and began to bite furiously at the netting. I took the beetle out of the mesh as it squeaked incessantly, overjoyed with this marvellous catch.

But what made it fall?

I couldn't quite put my finger on it, but something didn't feel right. I craned my neck up, and saw a paper mulberry tree about thirty feet up on the first tier of the precipice. *It fell from there.* I was sure of it, as I remembered that these beetles would 'feign dead' and drop out of mulberry trees when I used to catch them as a child. Something must have spooked it; besides, the timing seemed too much of a coincidence. I took a few steps back, and carefully scanned the rocky platform above.

I wished to heaven that I had never looked up.

Several hundred feet of sheer rock stretched up towards the sun, whose rays now beat down into the opening with a blinding

The creature looked down at me

intensity. Myriads of grasses and gnarled bushes were clinging onto the cliff face, apparently untroubled by the combination of blazing heat and powerful spray. Through the mist I saw a dark shape crouched under the mulberry tree, a long-limbed figure with its hands resting on the edge of the crag. As I shielded my eyes from the bright sunlight, the creature looked down at me and revealed its face – but what a dreadful face it was! A broad grin extended up towards the cheekbones, beyond which lay the eyes, if they could be called eyes – two lifeless, black craters sunk deep into the skull, swallowing up the light that was cast upon them. The complexion was of a most sickening livid grey, which made an unsettling contrast with the angular and well-proportioned features. As I uttered a cry of horror, the hideous creature opened its flailing limbs and leapt back into the thickets above, beyond my line of sight. A fist-sized rock came tumbling down towards me, only just catching a ledge on the way, and crashed into the plunge pool along with the roaring cascades.

I realised that I had already cowered into a ball, and was beginning to shake uncontrollably. I could barely grasp what had just happened, and in my pitiable state I tried to reassure myself that it was just a large monkey or a gibbon. But that horrid face! So distinctly human, yet so disturbing that the thought of it made my hair stand on end. I could find no natural explanation for it, and in a moment of clarity the strange circumstances leading up to my adventure finally hit home – the curious collection, Hakomoto's obscurity, Max's scepticism... nothing was as it seemed.

I suddenly felt like a helpless pawn caught amid a dark and twisted game.

ILLUSIONISTS

A CURIOUS COLLECTION

London, June 2017

Just a few weeks prior I had stood triumphant, wearing a smug smile as the great entomologist goggled at the collection which I had brought before him.

'That's very odd. Purple *Rhomborhina*! Local variant, I presume… Wow, those are some spectacular *Lucanus*! From the foothills of the Himalayas, surely?' Max exclaimed, as he cocked his head to read the tiny label. 'Who is this Hakomoto? I've never heard of him before!'

'Of course you haven't!' I laughed. 'He's not an entomologist as such, but an insect dealer. Makes his living through selling specimens. Mind you, they are of the highest quality, complete with collection data which would put most *real* entomologists to shame.'

'He certainly has a flair for collecting! Series of immaculate specimens, along with some remarkable colour variations…' Max's gaze was still transfixed on those purple cuboid-shaped beetles from the cardboard box, and he began to scrutinise one of their antennae with his hand lens.

As he performed this examination, I proceeded with my account. 'This box consists of Hakomoto's most prized species, collected from famous localities all over China. As you noted, there are colour morphs of flower chafers which I had no idea even existed, and the mandibular development on those bronze stag beetles is staggering – what's more, they're from Mêdog in Tibet! This collection must be worth a fortune!'

'I can imagine.' Max furrowed his brow in concentration, as something in the corner of the box caught his attention. 'The question is, how did this come into your possession?' He threw me a quick glance.

'You see, it's quite complicated.' I cleared my throat, and tried to recall the numerous emails that I had received earlier that year. 'He sent me these specimens to use as drawing references, since he had commissioned me to illustrate them. I told him that macro-stacked photographs of the individual species would suffice, but he said that he wanted the details correct to the last hair, and insisted on sending the original specimens over. When I finished the job, he surprised me by saying that I didn't need to send the collection back, and that I could keep it as my pay. It was a great relief for me, as I would have been very worried to send such valuable specimens back to Japan.'

'Was there a return address attached somewhere?' Max interrupted, stroking the torn faces of the cardboard box.

'No, it just came like this. Duct-taped all around with an address label on top. I didn't do a good job at removing them!'

I chuckled. 'But what an incredible deal! He never asked for the drawings themselves, only the corrected digital scans – and he said that I was free to sell all the originals, plus any prints derived from them. Now the artwork is all up on his website, and his specimens are selling better than ever.'

'Interesting!' Max scratched his beard, lifting a slender, sinister-looking beetle up by the pin, and read out the collection data on the label.

'Forgive me if I pronounce this wrong: Shri-Man-Siang, Hainan, China. That's the island in the South China Sea, if I'm not mistaken. 18°53'52.7"N 109°42'03.0"E, alt. 1,729 metres. Twenty-ninth of May 2014. Under bark of rotting *Castanopsis* trunk, col. Y. Hakomoto. But this is a *Carabus* – I never knew the genus even existed in Hainan!'

'Neither did I, until I saw this collection. I keyed it to *Carabus ignimitella*, but there is something about that pronotum and elytral sculpturing that's bothering me. Looks almost… *pustulifer*-like, if you ask me.'

His eyes widened and those pupils dilated for a split second, and I could have sworn that a fleeting look of revelation passed over his face. In an instant, he had reassumed his concentrated look and proceeded to drop a bombshell.

'I'm fairly certain this is a new species.'

Those words chimed like music to my ears. 'Well, why don't we publish it then?' I beamed with excitement.

'Evidence, my friend! A single specimen is rarely sufficient these days.' He smiled at me, and carefully placed the beetle back into the box. 'Aren't you off to Hong Kong to see your dad? Hainan's just across the strait. It would be great if you could get more of these!'

I had just finished the final exams of my degree – those three years of monotony and stagnation had made me sick to my stomach. The thought of adventure gave me a real kick.

'That would be incredible! My flight leaves tomorrow. I'll pack my equipment tonight and book my flight to Hainan next week.' I could feel myself trembling with exhilaration.

'Excellent! You know the drill: lots of torching at night, especially on the damper evenings. And head for the peaks if you can!'

THE PLOT THICKENS

Did I really see that? Am I going mad?

These questions tormented me as I dragged my heavy boots through the river back towards the village. Visions of that stark, lifeless face consumed my mind, and the poor visibility through the tangled thickets lining the banks of the river only fuelled my paranoia further. I dared not make any loud noises or movements, but the sound of crunching gravel seemed to drown out even the crashing torrents. I ended up running most of the downhill sections of the river, and returned to the village just before sunset; exhausted, frightened and confused, I shut myself in and slumped against the door.

After seeing that terrible face at the waterfall, my adventure took a most unexpected and sinister turn. I dared not stray far from the village over the next few days, only making short excursions to the vast clearing that I had discovered on my first night – the details of which I shall elaborate on later. But it soon became clear that the villagers were not as innocent as they seemed, and that many clues to the identity of the creature and *Carabus* lay hidden somewhere in this unassuming place. This suspicion arose from a string of inexplicable events that occurred in the village several days after my journey upriver, and it was obvious to me that there was some much deeper devilry at play.

I had finally motivated myself to clear out the mess from the cupboard, as I had just done my first batch of laundry and needed somewhere to hang up my clothes. Inside there was a very heavy box full of packaging material, which I moved onto the floor to inspect – it was already open at the top and when I tilted it, out rolled what seemed at first to be an ostrich egg. Thankfully it didn't crack, however, and simply gave a dull thud as it hit the bamboo floor. It was a large, ovoid object, which was about as long and as wide as my face. The outside was beautifully polished, and gleamed pearly white under the sunlight. I picked it up, and examined the tiny writing printed near the top.

'Philips, H36GW-1000/DX... *One thousand watts!* What kind of joke is this?' I exclaimed, realising that this was in fact a giant light bulb, and continued my rummage through the box. Eventually I managed to find two more such bulbs, a lamp holder, an extension lead, a fishing rod, a washing line, a ball of string, two white bed sheets and a portable petrol generator, which was responsible for much of the weight. As I took these seemingly random items out one by one and lined them up on the floor, I began to wonder what the meaning of all this could be – these things had to have something in common, surely?

I grabbed one of the light bulbs and ran over to my host's hut to interrogate her about the equipment's origins. Jin was sitting on the porch listening to the radio, and smiled at me as I stormed past him. The hut was empty. I hesitated, and remembered the old lady's reaction when she first saw me. *No, she had something to do with this.* I stepped back onto the porch, and placed a hand on Jin's shoulder.

Jin was sitting on the porch

'Tell me, Jin. Whose stuff is this?' I waved the light bulb in front of him, trembling with anxiety.

'What's the matter, sir?' The boy seemed quite taken aback by my reaction. 'These were left here by the person before you; you're not the first one to come here lookin' for bugs!'

I gulped. *It could only have been him.*

'How long ago? What did he look like?'

'Oh, must be a few years now! He was a strange man, sir, small but real stern-lookin', and barely spoke a word.' Jin lowered his voice to a whisper. 'And he was here with a white man.'

'A white man?' I echoed. 'Are you sure?'

'Yeah I swear, sir, the type you see in storybooks, with the big nose and everythin'. He was a very nice man, sir, like yourself. Left us a lot of money to keep his equipment safe and sound.' He nodded at the light bulb.

'Did he ask you to do anything with it?' A feeling of uneasiness was creeping up on me.

'Yeah, sir, as a matter of fact he did. I couldn't understand much of what he was sayin', but I think he wanted me to light the bulb and catch all the beetles that came to it. Just a second, sir.' Jin rushed back into the hut, and came back with a large spirit jar full to the brim with pickled insects. 'He gave us this, said he'd be back in a year to get it and pay us for what we caught. Never came back for it though, and I only managed to light the bulb a few times before the generator ran out of petrol. But I caught a lot of interesting beetles, sir!' He handed the jar over to me, with a proud look on his face.

Of course, this was a light trap! *How have I been so blind?* I thought to myself. *What else could a thousand-watt bulb have been used for in a place like this?*

I took the spirit jar, and examined it along with its contents with the utmost interest. The bottle was of the type used for storing chemicals, manufactured from borosilicate glass and fitted with a blue screw cap. I opened the bottle, and caught a whiff of the vile potion – unmistakably ethanol, but of a pure laboratory grade rather than the cheap antiseptic type. The black mass of specimens inside made it difficult to discern what exactly was there, but I could already see some unusual beetles, which I had never seen before. *Was it him? Could this be his method?* I could swear that my trail of thought was leading somewhere, if it weren't for the conflicting information in the boy's answer.

'*Never came back*? What about the small fellow... what were their names?' I pressed him for more details.

It was then that the old lady returned, with a basket of papayas she had just harvested from the orchard. She froze for a moment with her mouth agape as she saw the light bulb in my hand and realised that Jin had told me everything. Her face dropped; she grabbed the boy by the arm and rushed into the hut, slamming the door behind her and leaving me standing on the porch clutching the jar of dead insects.

That experience left a most disturbing impression on my mind, and raised far more questions than I could possibly hope to have answered. Why was the old lady so afraid? Who were those people? Why didn't they come back? My head was in a whirl, and the more I reflected on the matter, the more confused I became. In the end, I decided it was neither productive nor healthy to continue to stress about something I couldn't control, and instead focused my attention on the equipment that had passed into my possession.

I was very curious to try light trapping, as Max had always told me how wonderful it was as a collecting method and how it was the only way to collect many of the more elusive nocturnal insects – but the generator was completely empty. I asked around the village with the slimmest hope of finding petrol, but was amazed when one of the villagers showed me a jerrycan sitting in the corner of the tool shed which held about five gallons of petrol. He told me that it was used as an emergency backup for the only motorised vehicle in the village – a strange three-wheeler with the front of a cabin motorcycle and the rear of a pickup truck. The thing looked so dirty and beat-up that I wondered how it could carry any weight, let alone make it off the hillside in this sort of terrain. The man accepted my offer of 200 RMB for the whole can, which was more than enough to recoup the cost of transport and the fuel itself.

With the generator topped up and ready to go, I suddenly realised what a foolish decision I had just made. What if it didn't start? I might have just wasted a load of money on something I would have to decant right away. I flipped on the engine switch, and pulled the start cord. The generator coughed into life, and the splutter soon died down to a steady purr – what a sweet sound it was! Jin had kept the generator oiled and in prime condition, so that even after years of inaction it managed to start with no problems whatsoever. Now all I needed to do was find a spot to set up the light trap, and I already had the perfect place in mind.

In the few days following my journey upriver, I was so physically and mentally drained that I made no further attempt at scaling the peaks. The precipice at the waterfall was far too high to climb, and made even more treacherous by the spray of the cascades bordering it; besides, no force on earth could bring me back to that place, so disturbed was I by what I had seen. Instead, I decided to investigate the clearing that I had found on my first night, and see whether the path leading to it continued any further up the mountain. Sadly, the rocky gully beyond had become completely overgrown, and was so crowded with bamboo thickets and vines that only a bulldozer could have made it through. The clearing itself, on the other hand, proved to be a most pleasant place – the open space allowed me to see my surroundings and relax a little, although I still felt like I was being watched and maintained my vigilance at all times. I was very surprised at first to find how good the clearing was for collecting and spotting wildlife, seeming to far surpass the jungle itself in terms of diversity and abundance, but on second thought this made total sense – it was a mere illusion. The jungle is a multi-tiered ecosystem, with the bulk of its inhabitants living high up in the canopy; out of the minority that live on the gloomy forest floor, many are hidden deep within the leaf litter or extremely well camouflaged, like the leaf butterfly, which startled me on several occasions by 'coming to life' and taking off from a pile of shrivelled leaves. Even to the trained eye, most lifeforms in the jungle remain invisible and out of reach, requiring highly specific knowledge and methods to find – I would go as far as to suggest that less than 1 per cent of the insect species in the jungle could be found by 'casual searching', and that, furthermore, only a fraction of that minority might actually show up on the day. When the rainforest is cleared, the different tiers are squashed down into a virtually flat profile that, despite the massive losses in biodiversity, appears richer owing to its increased transparency.

The leaf butterfly

As the goal of light trapping is to attract as many nocturnal insects to the light as possible, the clearing occurred to me as the ideal location – the light was not obstructed by the canopy and could be seen from the skies all around, and any insect flying towards the source would not be impeded by the dense vegetation. I had all the equipment laid out and ready by late afternoon, but was confronted with the challenge of hauling a fifty-pound generator up the mountain for almost two miles. All I'll say is that I got my fair share of exercise on that day, and as I collapsed into a sweaty heap at the top of the trail I wondered why I kept doing this to myself – every time I promised myself rest after an exertion, I ended up getting carried away and doing something even more strenuous. But it was worth the effort. As I emerged into that familiar clearing, I was presented with a sight more beautiful than I could have ever imagined.

The sun was sinking behind the rolling hills in the distance, emitting a warm crimson glow that lit the clouds like roaring flames. I rested back on my heels, breathing slowly as I watched the dark clouds mingle with the purple mist rising from the mountains. I gazed over the horizon as the last rays of sunlight dwindled beneath it, and in that moment, the whole universe seemed to have shrunk into the view before me. As a soft chorus of crickets and frogs began to echo through the cool evening breeze, I gently closed my eyes, and felt like I was the last person left on earth.

BEACON OF THE NIGHT

Night falls quickly in the tropics, and within an hour the entire place was plunged into pitch blackness. During that time, I scouted around the edge of the vast clearing and found a perfect spot to set up the light trap – between two medium-sized trees about fifteen feet apart, which stood on a gentle slope at the far end of the field. The last leg of the challenge consisted of hauling the generator up there, unloading all the equipment and figuring out how the various items fitted together. After some trial and error, I ultimately deciphered the uses of the different bits and pieces: the washing line could be stretched between the trees, and be used to support the weight of the bed sheet; the string could be used to tie up the corners of the sheet, and then fastened to the trees to keep the fabric taut; the fishing rod could be used to support the light bulb, where the cord of the lamp holder threaded perfectly through the rings. The bulb simply screwed onto the lamp holder, and the extension lead could be used to keep the generator at a distance and reduce the noise levels. With the engine fired up and ready to go, I checked the various connections and closed the circuit. An electric discharge zoomed through the mercury vapour, emitting an eerie blue glow, and the lamp crackled into life.

A most unexpected guest

In just a few minutes, the bulb had heated up to white hot, and emitted a light so powerful that it was blinding to even glance at. A dozen moths began to circle the light trap, gradually spiralling in towards the source and bouncing off the scorching glass with plumes of fine scales; a myriad of tiny flies and wasps had already peppered the canvas, and I marvelled at the diversity of the minute beetles in particular as I vacuumed them up one by one using my aspirator.[15] Soon I realised that the other bed sheet could be laid on the ground underneath the vertical one, so that any of the smaller insects which fell would sit on the white sheet and remain conspicuous.

With the light trapping in full swing, I was joined by a most unexpected guest – a large toad, who waddled his way onto the bed sheet and began to snap up the various morsels which he found by his feet. I had a difficult time trying to keep the more precious insects away from his big mouth, while frantically bottling up the continuous stream of beetles arriving at the trap. Furthermore, I noticed that some of the less mobile insects often didn't make it to the sheet; they remained at the periphery of the pool of light, requiring me to constantly check the surrounding vegetation for any missed catches. In the end, I decided that enough was enough, and after coaxing the chubby amphibian into my net I gently lifted him off the floor, being careful not to trigger the toxin-secreting glands at the back of his head, and took him to a ditch about a hundred yards away. With the toad

15 An aspirator is, in effect, a tiny vacuum cleaner powered by the entomologist's lungs; it is used to collect insects too small or fragile to be handled. The usual design is a glass tube stoppered at both ends, with a tube going through each stopper – one serves as the collection tube, and the other as the mouthpiece. A gauze mesh is used to filter the mouthpiece – otherwise the entomologist might get an unexpected meal!

appropriately relocated, I came back to fetch my searchlight – I was planning on walking off to search the surrounding area, and leaving the trap running so that it might be replenished with newcomers when I got back.

I had barely taken ten steps when I heard a shrill that made my blood curdle. The rasping wail! It was nowhere near as melodic or as intense as the sound I had heard on the river, but it was distinctly similar and alarming nonetheless. I spun around to face the light trap, just in time to see a huge insect hurtling towards the lamp – it gave a dull thud as its bulk thumped against the glass, followed by a horrid screech as it flopped onto the vertical sheet, barely hanging on with only two of its claws. In an instant the mystery was solved, and I burst into a hearty laugh as I realised how dim-witted I had been. It was a cicada! I used to catch them by the dozen when I was a child, but I hadn't seen one for years and had clearly forgotten what they sounded like. I walked up to the thing, and picked it off from the sheet. It began to scream with a deafening harsh tone, which made examining it a most unpleasant auditory experience – extraordinarily, I later found that the insect's abdomen was almost entirely hollow, and acted solely as a sound box to amplify the noise. When held against the strong light, I could see straight through the abdomen, and apart from a pair of membranous structures (which were probably the sound-producing instruments) I could see no other organs in the cavity. I had no idea how such a design could be viable in the long term, and indeed that may be the reason for the brevity of these creatures' adult lives.

Although the appearance of the cicada made me realise what the baffling siren sounds could have been, I was sceptical that they were made by this particular species. For a start, the sirens were much louder, to the extent that I had never even contemplated the idea of an insect being the culprit; that said, I had read sources stating that some cicadas could produce sounds of over 100 decibels, which was approaching the pain threshold of human hearing. I certainly remembered the sirens being so overwhelming that they seemed to completely fill the atmosphere around me and reverberate deep into my heart. The cicada which I had found only made stuttering shrieks, perhaps because it was screaming for dear life as I held it by its sides, but I was certain that it couldn't produce anything as precise or as resonant as the sirens. Those cicadas were either much bigger than this already-huge insect, or had a much more efficient sound-box system – or both. Once again, I was presented with only a piece of the puzzle, and to this day I have never set eyes on the creature whose siren songs haunted me throughout my jungle adventures.

After finding my first cicada, plenty more appeared

Light trapping, as it turned out, was one of the most exhilarating experiences I've ever had – I was so absorbed by the whole process that all sense of time was lost, and I ended up staying at the clearing until the break of dawn. After finding my first cicada, plenty more appeared, along with some enormous butterfly-like moths, announcing their arrival with some rather ungainly crashes into the bed sheet. Deeper into the night, more and more bizarre creatures began to emerge from the darkness and throw themselves at the light, some of which were the stuff of nightmares.[16] By midnight there was barely an inch of space left on the sheet, and the entire surface was covered by a crawling mass of thousands of insects. Sifting through the mess and picking out the beetles was no easy business, as many of the more obscure ones could seem like wasps or flies to the untrained eye.

In just a few hours of light trapping, I collected more specimens than I had in the whole of the previous week – I could barely identify half of them to family level, let alone to genus or species, and I had every reason to believe that a good portion of my catch was new to science. You may think that I was overly optimistic, but the majority of the specimens I collected were tiny, drab-brown beetles, which are heavily understudied, and any mass-sampling method in the tropics is almost guaranteed to produce new species. That is not to say there weren't any 'substantial' beetles at the light trap; indeed, far from it – about half a dozen rhinoceros beetles and various species of stag beetles showed up, as well as a very large longhorn beetle. I knew most of these beetles to species level, however, so there wasn't much

16 One such monster, the dobsonfly, features in one of my later illustrations. There you can see it alongside a hornet and its nest, which should give you an idea of its immense size.

point in collecting them; I simply let them sit on the sheet until they flew away the next morning.

The highlight of the night showed up about an hour before dawn, just as the adrenalin rush was beginning to wear off and I was finally feeling sleepy. The alien creature had probably been on the sheet for some time, but had escaped my notice as it was hidden inside one of the folds of the bed sheet – it looked like a fusion between a mantis and a wasp, and had kaleidoscopic eyes that hypnotised me the more I looked into them. I immediately recognised it as a mantisfly, which had been top of my 'insects to see before I die' list since childhood. I marvelled at its resemblance to a mantis, and noted that its likeness to a red paper wasp served as a form of Batesian mimicry.[17] In fact, it is akin to neither of them, and its closest relatives are the familiar green lacewings, which can be found in any British garden. The adaptations of the mantisfly were remarkable, and anybody who mistook this insect for a small mantis could be forgiven – it snatched small flies and mosquitoes from the sheet using its spiny front legs, with a speed and ferocity to rival its larger patronym. This was a beautiful example of convergent evolution, where organisms from entirely different lineages had found a similar solution to a common problem. Unlike the mantis, however, mantisfly larvae bear absolutely no resemblance to the adults and live a different lifestyle entirely: they are translucent, maggot-like creatures that parasitise female spiders, first sucking the spider dry while it's still alive and later entering the egg sac to feast on the contents within. I find such behaviours horrific, but fascinating – the pain and terror that the victim must have suffered in order to produce

---

17 Batesian mimicry is where a harmless species has evolved to resemble a harmful species to deter predators.

the beautiful creature before me was almost too much to think on, but I was also amazed by how such intimate relationships managed to evolve in the first place. Surely it could find a better way, you might say – why does the mantisfly need to inflict so much suffering on another animal when it could subsist on something else or at least kill its victim in a humane fashion? This is where human logic fails, and we must put aside our prejudices to view nature impartially – it is perhaps worth reminding ourselves that nature cares not for compassion, justice, advancement or any of the human virtues that we revere, and that what we see in nature is merely what happens to 'work' at a given place and time, and will remain that way until a better solution presents itself.

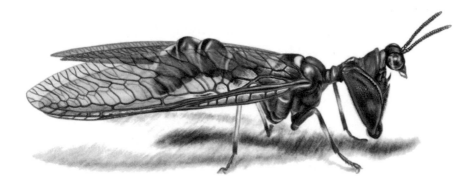

A fusion between a mantis and a wasp

I have often wondered why these creatures find light so irresistible, even when it can be fatal, and to this day there are few satisfactory explanations for this troubling behaviour. The situation is further complicated by the fact that insects attracted to light span a huge number of different families, and in all likelihood are not attracted for the same reason. The flight paths of moths towards light sources have been instructive in particular, as they have been shown to follow a precise logarithmic spiral. This surprising 'coincidence' is the result of their navigational method: to fly while maintaining a constant angle to the light source, which in a natural scenario would be the moon. When the light source is sufficiently far away, such as the moon, the flight path is as good as straight; when the light source is close, such as the lamp, the flight path becomes confused and the poor moth enters a gyre of doom. Not all insects spiral into light sources, however; indeed, many fly at them directly, and also seem to be less fussy about the wavelength of light emitted. Some describe this as the nocturnal creatures' attempt to find their way out of daytime hiding places, where the weak natural lighting has been overwhelmed by the intense artificial one – I can imagine this 'light-seeking' behaviour would also be very useful for escaping a confined space, say if they became trapped in a hole or crevice, which is a very real problem in the jungle undergrowth. In later conversations with a good friend, he noted that heat emitted from the high-wattage bulb may also play a role as one of the underlying attractants. Whatever the causes may be, we are still far from piecing together this complex puzzle – as with so many things in the natural world.

I should also mention that, much to my dismay, the toad somehow managed to find his way back to the light (or perhaps it was just another one, in which case they looked remarkably similar) and it became clear to me no matter how far away I dumped him, he was going to return with his insatiable appetite. On the subsequent nights of light trapping he showed up every time, and took full advantage of this free insect buffet – sitting thoughtfully under the light, picking off whatever took his fancy. It was often a plump cicada as the moths were covered with distasteful fluff and the beetles were a tough mouthful to swallow.

Towards the end of my trip, he had grown so fat from the feeding frenzies that I was worried about whether he could even walk. I have no doubt that the toad was extremely disappointed when, one night, I failed to show up at the clearing and he found all the equipment was gone – perhaps he is still waiting under those trees, wondering how those days of plenty came and went so abruptly, and dreaming of another entomologist to bring his next share of good fortune.

ASSASSINS

FATES OF THE INMATES

When dawn eventually broke and the first rays of light peered over the great mountains, I realised that I was faced with a most dreaded task. I glanced at the generator, and the thought of carrying it down the mountain alone was enough to make me weak at the knees. Even if I was able to, I would have had to carry it back up in the evening for the next light-trapping session – and I knew that I would never make it. I was willing to try anything to keep the whole set-up there for as long as possible, so in the end I took a leap of faith by leaving the equipment under a large tree, covered by a makeshift shelter constructed from dead branches and giant taro leaves. It was enough to keep out light rain, but should a thunderstorm hit in the afternoon then the generator would be as good as dead. Fortunately, the heavens never opened, and the rainforest didn't get so much as a drizzle; after sleeping the day away, I returned to the clearing with my waterproof tent in the evening, and my mind was finally put at ease.

In the days following that unforgettable night, I became almost as nocturnal as my quarry – I had realised that the efficiency of light trapping was second to none, and so I exploited this collection method to the max. This of course meant staying up late at night, sometimes till the next morning, as some unusual species of beetles could be trapped just before dawn. During the day, I would intersperse sleeping with sorting through the catch from the previous night, which took up a surprising amount of time. The spirit jar of insects that Jin had collected was interesting to sort through, although rather shambolic due to the boy's limited entomological knowledge – a hodgepodge of beetles, beetle lookalikes and anything the boy found remotely interesting. I couldn't help but chuckle as I picked out an entire hornet's nest from the jar, with a hornet and the larvae still inside. The mischievous child had clearly snipped the nest off its twig and pickled it right away! As the amount of alcohol in the jar was evaporating away day by day, I was forced to discard much of the bycatch in order to make space for the beetles I had collected – I kept the hornet's nest and a huge male dobsonfly, which I couldn't bear to throw away. As I had a wooden storage box and some insect pins to hand, I pinned up these wet specimens and left them to dry under the bed. In order to not confuse the different batches of beetles, I packed them separately into small resealable bags filled with alcohol, labelled with the collection data.

During these days I was so alternately excited and fatigued that I completely forgot about the enigmas that I had collected during my trip to the falls. This was perhaps for the better, as it saved me a lot of frustrating suspense: when I finally remembered them, the puzzles had already solved themselves. One of the spiky 'trilobite' pupae had already eclosed, and out from it emerged the most extraordinary

I kept the hornet's nest and a huge male dobsonfly

beetle I had ever seen – it was roughly the size and shape of a penny coin, and from the middle of the disc rose a conical dome of pure, blazing gold.[18] The translucent edges of its shell were drawn down over its entire body, so that the beetle looked like a miniature for da Vinci's prototype tank. I finally realised that this was the 'piece of gold' that had escaped right before my eyes beneath the falls, and now that I had the creature before me it was not difficult to understand why I had wondered if it was even real. The colour and shine of its dome were so convincing that I really wondered if

18   Eclosion is when an adult insect emerges from its pupal case.

underneath was a layer of gold leaf; furthermore, I could think of no survival value in this extravagance, since a bird could probably see it from a mile away. The beetle must have had some trick up its sleeve, so I put it on a leaf and began to observe its movements carefully.

For most of the time the 'golden tank beetle' sat completely still, but when it began to move the thing was comical beyond description – waddling along with large, snowshoe-like feet and wobbling its small head from side to side like a bobblehead toy. When I prodded it, the beetle tucked in its legs and pulled its tortoise shell tight against the leaf. I chuckled. *This is all you've got?* But when I tried to remove it, it became clear to me that I had severely underestimated the creature's powers – the strength of its grip was unbelievable, and the beetle seemed virtually fused to the leaf's surface. I failed to pick it up several times as its domed shell was extremely slippery, and even when I got my nails underneath the body to pry it off, there was some inexplicable force holding the thing down and it barely budged at all. In the end I had to take the leaf with the other hand and peel the two apart slowly so as to not hurt the animal, and I was left wondering if this was the reason why the flashy beetle hadn't been wiped out. Upon closer examination with the hand lens, I noticed that the undersides of its large feet were velvet-like in appearance – I had no doubt that they were a part of some clever underlying mechanism, and that there was even more to this beetle than met the eye.[19]

---

19 The underlying mechanism for this incredible strength is fascinating, and again explained succinctly in *For Love of Insects* by Thomas Eisner. These beetles are commonly known as 'tortoise beetles' and, in a nutshell, their strength comes from surface tension – just as two sheets of glass can be stuck together firmly by a drop of water, the tortoise beetle secretes an oil on its feet that forms a film of liquid between itself and the surface it's standing on. They can withstand pulling forces several hundred times their own body weight over a few seconds, and much longer if the forces are reduced.

The translucent blob with horns, however, didn't fare so well. On the afternoon it was collected it had already turned from a cool blue to a bright amber; at first, I thought this was a sign that the creature was about to shed its skin, but nothing eventful followed. The blob sat there like a statue – not moving, eating or showing any signs of life. After about a week, I noticed that a white object had appeared inside its body, and could be seen through the translucent walls. As the vials piled up in the corner of the room and I was preoccupied with sorting through the light-trap material, this great curiosity slipped completely from my mind. But a few days before my departure from the jungle, I suddenly heard a high-pitched buzzing noise coming from one of the vials, and on locating the one responsible I realised that the situation had only become stranger.

Perched on the leaf next to the blob was a small pretty wasp dressed in a gaudy contrast of black, white and orange like a fashionable model. The insect had a long tail-whip projecting from its rear, which I presumed to be its ovipositor. A gaping hole yawned on one side of the blob, from which the wasp had undoubtedly emerged. It was then that the truth dawned on me – this wasp was a parasite, as the blob was far too big to be its larva or chrysalis. The presence of an ovipositor is a characteristic feature of female parasitic wasps, which use it to pierce a victim's body and lay the eggs inside, thus confirming my suspicions. Having spent its entire life feeding off the internal organs of the unfortunate blob, the wasp had now burst forth from its host like a shocking scene from *Alien*.[20] It was a horrific but fascinating revelation, and explained many of the conundrums associated with the blob – the colour change to amber had been manipulated, presumably

20 Indeed, director Ridley Scott quoted parasitic insects as the main inspiration for the life cycle of the Xenomorph, the main antagonist in the *Alien* franchise.

chemically, by the wasp larva inside, and served as yet another form of aposematic warning against any curious intruders. The white object which I saw later was probably the wasp's cocoon, as I would've seen movement if it was still at the larval stage. It made my hair stand on end to think that the blob had been conscious all this time, paralysed and denied a painless death by the larva inside, which ate the vital organs last – this prevented its food supply from rotting, and, for all I knew, the poor thing was alive not long before the wasp flew off to find its next victim.

Like a shocking scene from **Alien**

It didn't fly far, however, as I grabbed the wasp and threw it into the spirit jar immediately. These specimens should always be collected, as they are extremely fortuitous to come by and provide invaluable records for improving our understanding of parasite–host relationships – indeed, I had stumbled upon a similar situation when I was younger, and with the help of Max and Gavin, the wasp expert at the museum, we jointly published the specimen as a new host record. After the creature had stopped struggling in the alcohol, I took it and sealed the specimen into its own packet.

'Gavin's going to absolutely love this,' I mused. 'And hopefully he'll be able to help me solve this mystery.'

THE PEAKS ARE BECKONING

This irregular nocturnal lifestyle, compounded by wild emotional fluctuations between euphoria and exhaustion, eventually took its toll – at the end of my second week I fell ill from a bout of fever that left me bedridden for three days. The jungle heat made me perspire profusely, and my aversion to the Hlai cuisine meant that by the time I recovered, I was so emaciated that I barely recognised myself in the mirror. During my suffering, Jin was kind enough to bring me lots of fruit and water twice a day. I knew how hard these things were to come by, so I made sure to repay him for his compassionate care; the boy refused any monetary rewards, but instead asked if I could tell him stories of my past adventures. He blinked nervously as I recalled a night-time encounter with a wild boar in France, and almost cried out with excitement when, at length, I managed to face off the beast with a kitchen knife.

'You're very brave, sir. But we have something bigger here – we have a Yeren on our mountains.'[21]

'*Ye-Yeren?*' I stammered, suddenly remembering a sight I had chosen to forget. 'Have you seen it before?'

'No, sir. But Grandma said it only lives on the peaks.' Jin stared at me sincerely. 'And it will eat anyone who gets up there.'

'Well, it would have a hard time sustaining itself, then!' I sniggered, realising that this was nothing more than an old wives' tale used to scare children. 'Don't listen to such gossip, Jin. Only believe something once you see it for yourself!'

Before my fever I had performed the dreaded task of bringing the light-trapping equipment back to the village, but thanks to

---

21 Yeren (literally 'wild man') is a legendary creature said to resemble both man and ape. It is the Chinese equivalent of the famous cryptid 'Bigfoot'.

the empty generator and downhill slope, the relocation was made relatively painlessly. Coincidentally, the monsoon downpours had also returned to the jungle, and I took solace in the idea that even if I had been well, I would've been confined to my hut anyway. Storm after storm ravaged the jungle, and bolts of lightning crashed down onto the mountainside in blazes of fury. With my mind cleared of such despondencies, I was lulled to sleep by the pitter-patter of raindrops every night, followed by a fall into deep slumber. This relaxation was no doubt the main reason for my speedy recovery – when the rains eventually ceased, I was fully recuperated and so invigorated from my comatose state that I set out for another expedition immediately.

From my previous ventures, I had established that the village trail and the jungle river were not feasible routes for the ascent – the former was clogged with thick vegetation beyond the clearing and completely inaccessible, while negotiating the treacherous precipices obstructing the latter was out of the question. I asked the villagers if they knew of some secret path, but, as expected, they never had a reason to climb to the summit and were even more clueless than I was. The only potential route that remained was the road by which I had arrived at the jungle, and it was my last hope of making it to the top – my last hope of finding *Carabus*.

I left the village early in the morning, in good time to avoid the afternoon thunderstorms; the monsoon season was now in full swing, and every afternoon the showers poured down like clockwork. As soon as the downpours ceased, the baking tropical sun would reappear to evaporate all the water back into the air, creating a misty wonderland that was rather more pleasant to look at than to explore. The humidity approached saturation point, so that even a temperature in the low thirties was enough to soak me in sweat from the slightest physical exertion. This general dampness, compounded with an abundance of mosquitoes and leeches, made the environment uninviting to say the least. Despite all this, I was adamant that I would conquer the peaks this time, and I set off down the mountain road in search of a viable path upwards.

It wasn't long before I reached the end of the tarmac, and the road turned into a rocky dirt track that stretched upwards for as far as the eye could see. Despite being obscured by trees, a break in the canopy was only just visible and seemed to be headed towards the summits on the western slope. My feelings were mixed: even though I was excited that I might have finally hit upon a gateway to the summit, I regretted not having explored this path earlier as it was the most obvious option. It would have saved me from the endless frustration of encountering the dead ends beyond the clearing and the waterfall. But here I was – blessed with a gentle, well-trodden trail that led me right to the misty peaks above, and finally a realistic chance of hunting down my quarry.

I was mistaken, however; even this apparently straightforward route turned out to be fraught with hazards. About a mile into the trail, the gentle slope suddenly rose into a huge escarpment, which was much steeper than it had appeared from afar – for most of the time, I was forced to scramble on all fours as my legs were giving out at the mere sight of the gradient. Signs of human activity gradually waned as I delved deeper into the jungle, and the path became more overgrown the higher I climbed. *What if this also turns out to be a dead end?* I couldn't get rid of the nagging feeling that this footpath used to be a game trail maintained by hunters, but had been abandoned long ago. To add to my misery, the ground was crawling with huge carpenter ants almost an inch long, whose trails criss-crossed the jungle floor to make the most of the open space. During the ascent I would often grab a prominent rock, and then cry out in agony as I accidentally crushed an ant under my fingers – yes, large ants can sting like wasps, and it is very painful indeed! After learning my lesson a few times, I became a lot more cautious about where I put my hands and would take a good look at all handholds before committing to them, a habit that would prove most valuable later that day.

Some of the ants, however, were so tormented that even I began to pity them, despite the pain they had inflicted on me. I first noticed this phenomenon when I was taking a water break by the side of the trail, and saw that an ant was clinging to the underside of a leaf with a most extraordinary outgrowth protruding from its head. When I examined it more closely, I saw that the ant was covered with a brown mould and was clearly dead – it seemed to have been infected by some horrendous disease, and at first, I considered the projection to be a kind of tumour. As I felt its leathery texture, however, I immediately realised that I had chanced upon one of the queerest organisms on the planet: it was a mushroom of the fabled 'zombie-ant fungus' that I had heard so much about in documentaries, whose life cycle was a testimony to how truth can really be stranger than fiction. A few weeks earlier, the worker ant had picked up a spore of the fungus while foraging, which then penetrated its exoskeleton using a combination of mechanical force and enzymes. Once inside the ant's circulatory system, the fungus began to secrete a neurotoxin that drastically manipulated the ant's behaviour, compelling it to climb up a plant and clamp its jaws around a vein on the underside of a leaf. The location chosen is always one that is optimal for further spore dispersal; once in position, the fungus killed the ant immediately and destroyed the fibres within the jaw muscles, effectively locking it in place with its own 'death grip'. The result was this strange scene that told of the inherent cruelty of nature, and I had no doubt that the spores from this mushroom were later released into the air to infect more unfortunate victims.

A relative of the zombie-ant fungus, the caterpillar fungus, lives on the Tibetan plateau and uses a similar method to parasitise the larvae of ghost moths. It has been known to the locals for centuries, and its Tibetan name *yartsa gunbu* literally means 'winter worm, summer grass', which reflects the people's wonderment at its paradoxical nature. The difference is that while the zombie-ant fungus remains relatively obscure, the caterpillar fungus is currently the most expensive mushroom in the world – it has been hailed as a wonder drug that supposedly cures diabetes, hypertension, impotence and cancer, among other things, and as a result is worth more than its weight in gold. Every year, hundreds of tonnes of the parasite are harvested from the Tibetan plateau, which are then sold to traditional Chinese medicine shops across Asia. As with many such traditional remedies, however, the purported effects are largely rooted in superstition and anecdote, with little to no scientific evidence to back them up.[22] Although recent studies have isolated from the mushroom the active ingredient cordycepin, a compound with potential anti-inflammatory and anti-tumour properties, it has now been

22 Such traditional remedies can occasionally be successful, but often because of the placebo effect rather than any active chemical present within the remedy. In a nutshell, a placebo is a drug that works only because the patient believes it works. I would highly recommend reading about some of the famous experiments regarding the placebo effect, since they demonstrate elegantly the power of the mind and how our perception of well-being can be easily distorted.

synthesised in the laboratory and trials involving the chemical have not lived up to its vastly exaggerated powers.

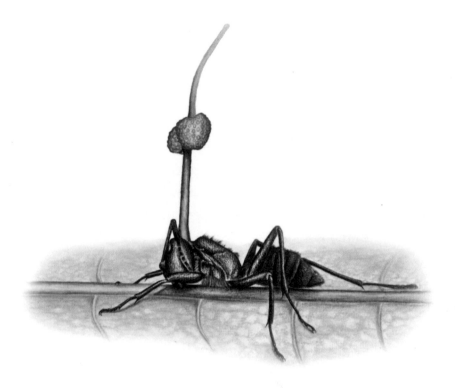

The fabled 'zombie-ant fungus'

Try as it might, the worst an ant could do was leave a few blood spots and a rather sore finger. There were far deadlier forces lurking in these forests, and the caution drilled into me by the ant stings became a lifesaver later on. The incident occurred at about 4,000 feet up, just as I was struggling up an exceptionally steep section. When I reached over to grab a branch to haul myself up, I instinctively recoiled as I glimpsed an unusual shape from the corner of my eye – there, curled up on the end of that very branch, a green pit viper poised and ready to strike! I dared not move an inch, and stood like a petrified statue waiting for the final blow. The snake gave an ominous hiss, and began flicking its forked tongue rapidly. That terrifying moment was forever burned into my memory – as I recovered from the shock I gradually edged my way down the rock face, cowering like a mouse before an angry serpent and praying that I hadn't startled it enough for an attack.

After what seemed like an eternity, the snake relaxed its guard and slithered back into the thickets, leaving me paralysed with terror and wondering if I could push my luck any further. Although this snake was not the most lethal of its kind, it could have easily killed me with a single bite – I was over half a day away from the nearest hospital, assuming that I could even make it off the mountain on my own, and at best I would have become an amputee.[23] It was only upon reflection that I realised the sheer severity of the encounter, and how every single one of my actions and decisions had potentially dire consequences. But no matter how vigilant I was, the ascent was becoming more treacherous the higher I clambered: in places, the thick lowland rainforest thinned out into growths of stunted trees and shrubs that clung onto crag after crag of granite slabs and ferrous outcrops. As the weathered surface rocks began to crumble away beneath my feet, I was forced to use the trunks and stems as handholds – these plants had root systems extending deep into the rock face, which dissipated any pressure placed upon them and were less likely to break. There were no guarantees, however, and the climb was as much of a mental challenge as a physical one, despite

---

23 A much larger, and more dangerous, species of pit viper exists on the mainland. It is known locally as the 'five-pace snake' as it is said that after being bitten one cannot make it further than five paces alive. This is, of course, a vast exaggeration, but the snake does indeed have a very potent venom and can kill within hours.

my years of rock-climbing experience. Eventually I topped out onto a ledge at 5,000 feet, which extended back into a platform marking the end of the steep slope. Trembling from the stressful exertions, I decided it was finally time to take a break and get some nourishment.

A green pit viper poised and ready to strike

A problem I had in the rainforest was that no matter how exhausted I was, I always found it very difficult to stay still. Perhaps it was because there was so much to see and explore, or the fact that every time I made a serendipitous discovery I would think: *Thank goodness I took the effort to get up and look!* The joy which this process brought me was ineffable, and a 'good catch' often left me in a state of ecstasy for hours on end, shaking with excitement as I stared blankly into the vial. You may laugh at such childish self-satisfaction, but do we not all yearn for these moments? So instead of sitting down, I wandered off for a leisurely stroll, sipping cold spring water I had collected from a stream earlier, and munching on a box of plantain chips I had fried up in the morning. *If the weather was brighter, this would be the perfect picnic spot,* I mused to myself. Through the fog I could only just make out the forms of the foothills below, and I imagined that on a sunny day this lookout would have offered a spectacular view of the lowlands. Regrettably, there wasn't much to be seen, so I walked further up the platform towards a grove of trees at the far end.

For once, I wish I had stayed still.

A giant dead tree slumped in the corner, head down so that its fragmented crown was teetering on the edge of the cliff. It had crashed out of the dense jungle growing on the gentler slopes above, presumably brought down by a violent typhoon that ended abruptly its long and peaceful life. Judging from the stage of decay it had been deceased for many years, as the bark was now peeling away from the rotting wood inside. I began to dig through the soft sapwood with my hands, and found some beetle larvae almost immediately – a dead tree is truly an entomologist's dream! Just as I was bagging them up and planning on digging for more, my gaze fell on a curious mark that tailed off as it curved around the trunk.

What on earth is that? I climbed over the log, and saw that the trunk on the opposite side was covered with numerous lacerations. Some of them were single deep gashes, others were multiple shallow cuts made from different angles, as if wedges of wood had been removed by a mechanical device. Shredded bark, pieces of sapwood and heartwood were strewn all over the ground, so that the horrific scene looked like the disembowelled corpse of some ancient giant. It was difficult to tell how old these marks and debris were, as they were all covered with a thin layer of green algae from the general dampness, but the edges remained sharp and well defined. They could not have been made by an animal, and it definitely wasn't a chainsaw. As I looked down at the destruction which the attacker had left in his wake, the truth suddenly hit me.

These were machete marks, made with the sole purpose of dismembering the log to reach whatever was inside. But what was he after? The same things I was after, presumably. I gulped. *It was him. He never left.*

LABYRINTH OF ROOTS

I took a crumpled piece of paper from my pocket.

18°53'52.7"N 109°42'03.0"E, alt. 1729m.

The location was about half a mile due east of me, and another few hundred feet above. All I needed to do was continue straight up the western slope, and skirt around the sheer cliff just visible through the mist. The terrain beyond the platform had flattened out significantly, and the vegetation here returned to its natural luxuriant state. If anything, this high-altitude forest seemed even more complex than the jungles below, perhaps due to the myriads of epiphytes covering every inch of visible bark − lichens, mosses and bromeliads of the most fantastic shapes hung like coral necklaces from the gnarled branches, upon which the moisture collected and ran off in continuous streams of condensation.[24] It is for this reason that these habitats are classified as 'cloud forests' − cold, mysterious

---

24 Epiphytes are aerial plants that grow on other trees but do not adversely affect them − such plants have left the soil completely and derive all their sustenance from air and rainwater. Some epiphytes, such as bromeliads, can also accumulate debris at their base, thus providing another source of vital nutrients.

and covered in perpetual fog. Such forests can only grow near the peaks of high mountains in the tropics, and the cloud forest's sudden appearance was a promising sign that I was on the home straight. But just as I believed I was set on conquering the summit, the mountain had a final trick up its sleeve.

Almost as soon as I entered the cloud forest, I was confronted with a decrepit door blocking the path. The entrance was flanked by wire fences, which extended far into the forest on either side and disappeared into the mists beyond. A thousand thoughts ran through my mind. *Who built this? Why in such a place? What if this is a restricted area?* The Chinese are infamous for placing secret military units and operations inside remote mountains, and they wouldn't hesitate to jail any intruders – no matter where they came from or what their intentions were. I moved closer, and saw that the only thing holding the door closed was a rusty padlocked chain that dangled ungainly from a gaping hole drilled through the board. There were no warning signs of any kind on the barrier, and its poor condition told me that it was no longer in use. At least it seemed that way.

I peeped through the hole in the door. Under the dim light of the jungle undergrowth I could see nothing but a tangle of tree roots, which extended up the hillside in a ladder formation for at least a hundred yards before it curved out of sight. 'Strange!' I muttered out loud. *Why are these root systems exposed? It would only take a few storms to cover them up with sediment again. Unless…*

This is the path! I suddenly realised. This underground network had been kept exposed by travellers of the past, and the unnatural polish on some of the larger structural roots indicated that they had been repeatedly used as handholds. There were no footprints visible in the mud, but the lack of vegetation on the ladder hinted at a usage frequency far higher than its remote location would suggest. *This is where the fabled Yeren comes from*, I thought. The gnarled, intertwining

shapes of the giant roots reminded me of the Minotaur's labyrinth –
indeed, the fence and padlock made me wonder if there was some
monstrosity which the constructors were trying to confine.[25]

With a great pull, I hauled myself on top of the door and
vaulted over it, making a quick breach into the forbidden zone.

The entrance was flanked by wire fences

For some reason, there was an eerie silence. The birds in
the forest had stopped singing, and all I could hear was the soft
dripping sound of water onto the leaf litter. I sniffed nervously.
There was a funny scent in the air, which tasted almost metallic
in my mouth. Maybe something had spooked the birds, or maybe
they had sensed something beyond my perception. I tried to ignore
the unnatural stillness, and began my ascent of the root ladder.

25 In Greek mythology, the Minotaur was a half-man, half-bull monster who lived
    in a labyrinth and ate anyone who became lost there.

It was not long before I reached my destination. Ironically, the thing which I called 'the labyrinth' proved to be a great asset during the last leg of my journey, as the roots were extremely sturdy and reliable as handholds. If it weren't for this clear path, I would surely have become disoriented – the fog was so thick up here that I could barely see twenty yards ahead of me. The shadowy forms of twisted trees stood like nebulous sculptures dissolving into the mist, and the general feel of the whole place was one of haze and paranoia – to the extent that I had no clue of my surroundings or what time of day it was. Then again, I wasn't sure I wanted to know. The only thing I trusted was my altimeter, and as the coordinates aligned, I finally emerged onto another platform at the altitude of 1,729 m.

This was it. I had often tried to envisage what this place would look like, and it was exactly how I had imagined it – damp, alien and bleak, this rocky precipice was a far cry from the steamy jungles below. Despite what it seemed like, this was the perfect habitat for *Carabus*, as it was the only place with a temperature that the giant ground beetle could tolerate. Since Hainan was still connected to Indochina during the last Ice Age, it was possible that the flightless *Carabus* had spread across to the island over the land bridge. As the global temperatures soared, however, the temperate woodlands rapidly became tropical rainforests, and it was this abrupt change in climate that had pushed these colonists to the peaks of the highest mountains, where the last of them remained. Considering that these beetles had been isolated from the mainland populations for at least 10,000 years, and that their new environment presented unique challenges, surely the ones here had already evolved into a new species? There was only one way to find out, and that was to catch one. Now I was up here, I finally had the chance. I took out the crumpled piece of paper again.

Under bark of rotting Castanopsis *trunk.*

The final step was to find a living *Castanopsis*, of which there were many. The trees were easily recognised by their rough, oak-like bark and the male trees were also sporting upright catkins, which grouped into spectacular white clusters resembling exploding fireworks. None of them grew beyond about twenty feet, presumably due to the poor soil quality and high wind velocity, and their great age twisted those thick trunks into the most fantastic shapes. Among them I spotted a particularly stunted one with a rotting limb, and was extremely surprised to find a *Nepenthes* pitcher plant sprouting from it, which I didn't even know existed in China. I looked closer, and suddenly realised that the 'pitchers' were in fact extensions of leaf veins, as the youngest ones looked merely like tendrils with slightly distended apices. *Nepenthes* plants have been known to grow into shrubs that aggregate to cover entire hillsides, but the one before me was still in its infancy — a few of the pitchers had withered, others had not yet opened and only a couple were functioning traps. I had long heard about their carnivorous behaviour, and how they used their pitchers to trap their prey, but never understood how they achieved this. It seemed strange to me that any insect would climb into the pitchers spontaneously; besides, the lids of the contraptions were fixed in place and couldn't close to trap what was inside, so why didn't the inmates escape? My question was soon answered by a real-life demonstration. A carpenter ant crawled around the rim of a medium-sized pitcher. It headed straight for the underside of the lid, and began to lick its base; then, as soon as the ant tried to climb onto the lid, it lost its footing and dropped like a stone. I looked into the pitcher, and saw that it contained a cloudy liquid — presumably full of digestive enzymes and antibacterial agents. It became clear to me that the lid served as a 'roof' to keep out rain, which would not only dilute the digestive fluid but also overfill the trap and render it useless. On the liquid surface floated about a dozen other dead ants, which had probably also been attracted and fallen to their demise for the same reason.

A Nepenthes pitcher plant

But what's so irresistible about the pitcher? I sniffed the plant, but it didn't emit any detectable aroma. I then touched the base of the lid, and found it to be covered with a sticky liquid –

97

being a typical naturalist, I proceeded to lick my finger and was rewarded with a droplet of sweet nectar. *What can this lure?* I checked the other two pitchers: the small one had only a few ants, and the large one... a living beetle, still struggling in the liquid! It was a ground beetle – that much was certain – but it was a very strange and slender species unlike anything I had seen before. I reached a finger inside the pitcher to let it crawl onto my hand, and as the beetle rapidly climbed up my arm a shot of pure adrenalin rushed through my veins. *This is it!*

I grabbed the beetle, and held it firmly between my fingers. The creature began to wave its legs around frantically and attempted to turn around and bite me, but any attempt to escape was futile. *I've got you,* Carabus, *once and for all!* It was a beautiful specimen – sculptured like an emerald figurine, and glowing with a brilliant metallic sheen. Furthermore, it was clearly distinct from the much larger, but much drabber specimen in my collection. I held it closer to examine its glistening shell, and was already dreaming of what name I would give to this new species when...

The beetle extended its abdomen, and flicked up a fine mist of acid.

It all happened too quickly. I instinctively flinched, but my eyes were already burning with a scorching sensation. I choked as a most horrendous smell filled my nostrils and the acid seeped slowly into my eyes. In the confusion I must have dropped the beetle to reach for my water bottle, as I desperately tried to rinse the vile chemical from my eyes. The pain was excruciating – for a moment I thought I was definitely blinded, but as I slowly opened my swollen eyelids I could just about see the blurred outlines of my palms. *And they were empty!* In my agony and sorrow I remembered an anecdote about the great naturalist Charles Darwin, which I used to always tell my friends as a joke. With my recent misfortune, however, I now sympathised with him deeply. Below is an account of the incident, in the man's own words.

A *Cychrus rostratus* once squirted into my eyes and gave me extreme pain; and I must tell you what happened to me on the banks of the Cam, in my early entomological days: under a piece of bark I found two *Carabi* (I forget which), and caught one in each hand, when lo and behold I saw a sacred *Panagaeus cruxmajor!*[26] I could not bear to give up either of my *Carabi*, and to lose *Panagaeus* was out of the question; so that in despair I gently seized one of the *Carabi* between my teeth, when to my unspeakable disgust and pain the little inconsiderate beast squirted his acid down my throat, and I lost both *Carabi* and *Panagaeus!*[27]

Wallowing in my own self-pity, however, was not going to bring the beetle back. As soon as my vision was restored I scoured the ground and leaf litter around the tree, but the little devil was nowhere to be found. To this day I have no idea how it 'knew' when to spray and where my eyes were, or how it managed to fire its chemical weapon with such precision. All I knew was that I had let a potential new species slip through my fingers, and that it was highly unlikely I would find another one. Not by conventional methods, at least.

The strange setting in which I had found the beetle got me thinking: *If the pitcher plant could trap one, then surely I could as well?* I

26   *Carabi* is the plural of *Carabus*.
27   This extract is taken from *Life and Letters of Charles Darwin* (1887).

dug into the depths of my rucksack, and pulled out a few plastic cups that I had kept from my plane journey. At the time this was simply an act of frugality, but now it was a blessing. I suddenly realised that the trick to successfully capturing *Carabus* lay in imitating the plant's methods. I set up a 'Barber trap', which entomologists use to catch ground beetles. The trap consists of a cup sunk into the ground, rim flush with the soil, and baited so that it acts as a sort of 'pitfall' for any curious insects to fall into. The cup is usually filled with preservative fluid, which instantly pickles anything that falls into the trap – but besides having no preservative fluid, I was not keen on the idea of killing so many other insects unnecessarily, as I was only after beetles. I cut some drainage holes in the bottom of the cup with my pencil knife, and placed a curved piece of bark over the trap as the roof; this way, the trap wouldn't become flooded even in heavy rain. A small slice of banana was placed in each trap as bait, as it was the only food that remained in my rucksack.

I had just finished setting the first one and was about to move off to set the next one, when I heard a familiar sound echoing up the mountainside.

It was the sirens, but this time from a solitary cicada. I wasn't sure whether it was due to the colder temperature or the creature being on its last legs, but the song sounded distinctly weaker – each note was a subdued hum, repeated eight times before the standard refrain of a long, rasping wail. It was coming from the forests below. I walked over to the edge of the platform, and looked down at the steep rock face. A thick blanket of fog covered the slopes, and the platform of granite upon which I was standing rose like a podium above the cloud forests below. As the ominous solo continued in the background, the chilling truth suddenly dawned on me.

This is the cliff. The driver's words came back to me. *The unidentified skull. The entomologists who never came back.*

Everything fit together neatly like a jigsaw puzzle, and I would have cried with alarm at this shocking revelation if it weren't for a distant crunch of fallen leaves, which stopped me dead in my tracks. For a moment I thought it was just a figment of my overactive imagination, but then there came a distinct rattle from the bushes behind me. I shivered, and turned abruptly to face my adversary.

The Castanopsis grove stood in silence

But there was nobody. The *Castanopsis* grove stood in silence except for the gentle rustle of leaves and the drip of condensation as it fell from the wind. I moved to another tree, and began to dig a hole underneath it for my next pitfall trap.

A cold, sinewy hand rested on my shoulder.

RELICS

YEREN

I closed my eyes. *This is the end*, I thought.

But the blow never came. I froze for a second from pure shock, and then my survival instinct kicked in. With a burst of energy, I grabbed the arm with all my might and threw my assailant over my shoulder. The creature hit the ground with a thump, and I found that dreadful livid face inches away from mine – I instantly grabbed hold of the head and dug my thumbs into the deep eye sockets but, to my surprise, they were completely hollow. A flimsy wooden mask peeled off to reveal the face of a giggling child or, more specifically, a giggling Jin.

I slumped to the ground, completely speechless. The boy lay flat on his back and couldn't stop giggling, as if this was all just a massive joke to him.

'You almost scared me to death, Jin!' I roared with anger, but instantly regretted my heavy-handedness. 'I hope you're not hurt? But why the hell have you been following me?'

'I was curious, sir. I wanted to see what your adventures were like.' He got up slowly, and patted the dirt off his trousers. 'And maybe learn some of your methods. Awfully clever, sir, the roofin' and the drainage holes.'

'Why all this, then?' I waved the mask in the air agitatedly, still struggling to believe that the horrid creature was Jin all along. 'You could've just asked me if you wanted to come along!'

'It's Grandma.' Jin lowered his head. 'She would never let me play on the mountains, sir, and would be worried even if you took me. So I wore this and followed you. I didn't want to be recognised.'

'Well, she's absolutely right! I barely made it up here in one piece myself, and look at me now!' I stood up and rubbed my swollen eyes. If I could have seen myself in the mirror, I would've looked like a broken runaway. 'This is not a place for games. What the hell do you think you're doing here?'

'I was only tryin' to protect you, sir.' The boy looked up at me with his large, innocent eyes. 'Lots of bad people in the mountains, plus the Yeren at the top.'

'*Do I look like I need protection?*' I scolded, although the sentence trailed off into a smile as I began to appreciate the boy's goodwill. 'You'll turn into a Yeren soon, with all that crouching and pouncing! Come on, help me set up these traps and we'll go home.'

The boy nodded, and began to dig away at the humus. As he hunched over the ground, I couldn't help but wonder at my own stupidity. *Of course it was Jin!* I remembered that there was something disturbingly familiar in the way the creature stooped over that ledge, which I should have instantly recognised as Jin's bad posture. Besides, there was no way that one of those entomologists could have lived as a hermit up here. I wagered that nobody could survive in these cloud forests for more than three weeks at a time, let alone three years – the frequent thunderstorms, lack of food and cold nights made the place completely inhospitable. As we finished setting up the last few traps, the pitter-patter of raindrops sounded from the canopy and we heard the unmistakable rumbling of distant thunder. A storm was coming, and there was no time to linger – I had originally wanted to climb to the summit, but topping out at this time would be as good as suicide. I did not want to become the lightning conductor for the whole mountain!

We raced along the labyrinth of roots, but it was already too late. A gust of cold wind howled through the trees, and torrential rain poured down from the skies. The stunted canopy of the cloud forest was useless for shielding us from the deluge, and the gale tore through the rainstorm to form a fine spray of water droplets, only worsening the already dreadful visibility. Within minutes we were drenched from head to toe, and gave up rushing as we realised that it was simply a waste of energy. Giant earthworms began to emerge from the forest leaf litter to slither over the soggy ground, their two-foot-long bodies glowing with an eerie blue iridescence. It was an alien sight, one which most would regard as their worst nightmare – but Jin looked happier than ever. He ran around poking and examining the soft-bodied creatures, and opened his arms to the sky as if to let the rain wash away his worries. I kept my head down and trudged through the mud, still filled with self-loathing over the fact that my complacency had cost me my prize. *What will I say to Max?* My last hope now lay in those pitfall traps, and I prayed that they would be filled by more than just ants when I next returned.

The giant dead tree lying at the edge of the platform

By the time we reached the end of the labyrinth, the storm had sapped me of all my energy. All I wanted was to get back to the village, have a hot meal and sleep the day away, but my body refused to budge. I flopped onto the platform where I'd taken my lunch break, and told Jin to rest as well before the steep section below. The storm had now by and large subsided, but there was still a steady drizzle that ensured that our clothes remained saturated with water. As I made a futile attempt to wring out my clothes and shed some of the weight, Jin spotted the giant dead tree lying at the edge of the platform and ran over to investigate.

'Save yourself some energy, Jin! You're going to need it for the next section. Besides, I've turned that place upside down; there's nothing left to find,' I protested, as he began digging at the pile of debris under the tree.

The boy gave no answer, completely absorbed in the wood pile. I was amazed not by his energy, but by how a superficially timid and introverted child was so full of passion and courage. It seemed to me that he had no fear for the elements or the harsh conditions around him, and that this was all just a treasure hunt to him instead of a dangerous expedition. After about twenty minutes, Jin got up and waved at me.

'Come over, sir, I think you'll like this.'

'Oh, what is it now?' I groaned impatiently, and staggered to my feet. As the blood struggled to reach my head, I nearly passed out from a severe bout of dizziness. I was far from recovered, I realised.

If getting up wasn't enough to make me faint, then the sight of the thing definitely was. The boy was holding a piece of rock, upon which perched what I immediately recognised as a 'trilobite larva'.

'Eureka! Eureka!' I shouted, and began dancing around like a lunatic. Jin raised his eyebrows and stared at me with a concerned expression, unsure of how to react to my manic outburst.

'What is it, sir? It looks awfully strange!' The boy prodded the creature, which reacted by shrinking its head back under its shell like a tortoise.

'Strange' was quite an understatement. I wouldn't have been surprised if these used to crawl on the bottom of Devonian seas, but to find such a thing under a log in a present-day forest seemed out of this world for me.[28] Look at its armour plating, its tiny snail-like head – how unbelievably bizarre this creature was! Only a few

---

28 The Devonian was a geologic period around 400 million years ago, when trilobites roamed the oceans. This hugely successful class of arthropods declined sharply in the Late Devonian epoch, and was eventually wiped out in the Great Permian Extinction around 252 million years ago.

months earlier I had read about the fascinating life cycle of this 'trilobite larva', which, believe it or not, was actually a *female beetle*. In a phenomenon known as neoteny, this animal would remain in its immature form for its entire life – it would never enter a pupal stage, nor would it ever moult into a recognisable beetle. As a result, I could not tell if this beetle was indeed a larva or an adult, but it was definitely a female – the males are minuscule in comparison and look a lot more like your conventional beetle, but are even more obscure and have been recorded only a handful of times in history.[29]

'Well done, Jin! You've found quite a treasure this time!' I exclaimed. 'This is a very rare beetle, and it would make a fine addition to the museum's collection.'

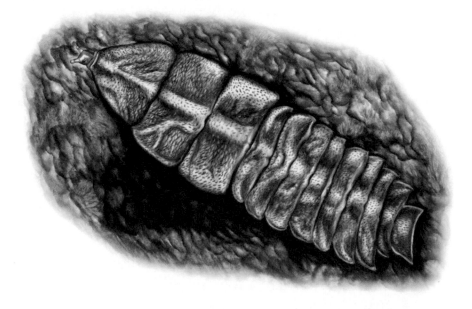

'Eureka! Eureka!' I shouted

29 The first male was discovered in 1925, about a century after the females. It was found by using a tethered female as the lure, and 'caught in the act' so its identity could be confirmed. It was about a tenth of the female's size, and died almost immediately after mating. Talk about matriarchy!

I decided not to elaborate on the identity of the creature, as I did not feel like Jin would appreciate its full significance. To a child, such things were better left as puzzles rather than facts, and merely as something to be pondered over – perhaps one day he will remember the moment when he discovered this little gem, and, curious to learn more, will return to discover the truth for himself.

FLYING DRAGONS

In the days following my turbulent trip to the peaks, I was forced to take a break from any strenuous adventures. My eyes were still feeling sore the day after the beetle sprayed me, and whenever I held anything in my hands my arms would shake involuntarily from sheer muscle fatigue. As I came to the realisation that there were only three days left of my expedition, I remembered the absurd reason I was forced to leave.

'That stupid graduation ceremony!' I muttered. 'What a waste of time! If it weren't for that buffoonery, I could have stayed here for another month at least!'

There was little I could do except to lounge around the lowlands and do some light collecting work. Jin acted as my assistant, and helped to carry the net around as I went about flipping over logs and searching under stones. After the discovery of the trilobite larva I began to pay much closer attention to detail, and instead of rushing around blindly trying to cover as much ground as possible, I decided to take a more thorough and diligent approach. The mindfulness paid off. Within the next few days I collected many series of rare saproxylic species – beetles that rely on dead wood for their development, and another community that is severely under-recorded. As I performed the meticulous task of disassembling the rotting trunks with a machete and sifting through the debris, Jin enjoyed himself by running about and catching whatever happened to be flying around. The boy soon became very adept at using the butterfly net. Initially he brought back a slow-fluttering damselfly, then he progressed to catching tiger beetles and carpenter bees, until later in the day he proudly took out a large dragonfly from the net, which I had no idea how he managed to capture.

I was very surprised, therefore, when one time he returned empty-handed, but beaming with excitement.

'I saw a dragon, sir!' Jin exclaimed, gesturing wildly with the net. 'A baby one! It was too high for me to reach, and flew away when I tried to catch it!'

I laughed incredulously. 'Don't be ridiculous, Jin! Dragons aren't real. You've been reading too many storybooks.'

'No, I swear, sir! It was just on that tree over there! Come on, I'll show you!'

He didn't need to take me anywhere. A flash of colour on one of the trees made me look up, and I saw a yellow triangle flickering up and down. It seemed as if the tree had suddenly come to life and began to wave a little flag, but I really couldn't see what on earth was waving it. Then, in a blink of an eye, a scrawny lizard leapt off the tree and began gliding – yes, *gliding*! With the backdrop of tree ferns and bromeliads, I felt as if I had stepped out of a time machine and into a lost world of flying *Microraptors*.[30] I stood with my mouth agape, hardly believing what I was seeing. It *was* a dragon.

---

30 *Microraptor* was a small, feathered dinosaur that had four wings and glided (may even have flown) between trees. It provided key evidence for establishing the evolutionary relationship between dinosaurs and birds, which is now widely accepted in the scientific community.

'Get it, Jin!'

I had no idea how he responded so quickly, but the boy must have anticipated its leap. He waved the net at the creature head-on, which crashed to an abrupt halt as it glided into the back of the mesh. For a moment I thought it was going to escape, as the reptile turned and attempted to jump through the mouth of the net, but Jin reacted immediately, and with a quick twist of the handle the netting folded over the rim so that the lizard became hopelessly trapped at the bottom.

'I got it, sir, I got it!' Jin dropped the handle and dived at the netting. He had caught the dragon.

'Those were some seriously impressive reflexes, Jin!' I gasped at the boy's sudden burst of energy. 'Watch out though, it might bite you!'

Jin carefully took the lizard out of the net, and handed it over to me. I marvelled at its extraordinary appearance. The reptile's back was covered with a lichen-like patterning, which explained why it was invisible when perched on that mottled tree trunk; furthermore, it possessed large lateral flanges which gave it a flattened appearance, so that when it pressed its belly against the bark it didn't even cast an obvious shadow. I gently pulled at the flanges that extended like parachutes into two huge flaps of skin growing from its sides, supported by immensely elongated ribs which could be collapsed whenever the 'wingsuit' wasn't in use. A strange, bony spike protruded from underneath its throat, which also supported a flap of bright yellow skin that wasn't visible when the animal was at rest: this was the 'flag' that I saw, and was presumably used both for display and horizontal stabilisation while gliding. All in all, it was a truly remarkable creature, and of a surprisingly gentle disposition – not once did the dragon attempt to bite me during the handling. When I released it onto a leaf it didn't attempt to escape, and just perched there looking at us with a funny blank expression.

The lizard spread its wingsuit and soared through the jungle

'Why isn't it runnin' away, sir?' Jin seemed puzzled at the creature's placidity.

'It's the way of reptiles, Jin!' I chuckled, and petted the lizard on the head. 'Lazy as a sloth for most of the day, and fast as a lightning bolt when it needs to be!'

Those words had barely left my mouth when the reptile darted up an adjacent tree; with a graceful leap, the lizard spread its wingsuit and soared through the jungle like a majestic dragon, disappearing as its form dissolved into the collage of greenery.

Following the release of the first one, we soon realised that these lizards were in fact very common. They were easy to spot with their ostentatious throat flags, which seemed to completely defeat the purpose of their brilliant camouflage – but I guess they had the capacity to take some risks, given their remarkable abilities. The males were constantly sprinting around on the trunks, displaying to the females while battling with the other males over territory. During this observation, I noticed that they particularly liked hanging around jackfruit trees, which seemed to be bustling with activity and causing the most conflicts. I did not think that the lizards were after the fruits themselves, but rather the flies and insects that were attracted to the rotting mess underneath the trees – either way, Jin and I decided it was a good place to stop for lunch, and make the most of this tropical fruit buffet.

After a trip to the cold and unforgiving cloud forests on the peaks, I truly began to appreciate the lowland jungle for its richness and fecundity. It was a place where, given the right knowledge and expertise, one could easily survive for weeks on end by living off the various fruits borne by rainforest trees. All the native lychees, figs and longans paled in comparison to the alien jackfruit – by far the most delicious, productive and wholesome of all the rainforest fruits, which often grow to an immense size of over two feet long and can weigh over fifty pounds. The golden flesh of ripe jackfruits is intensely sweet, and possesses a most peculiar but pleasant aroma – not as overpowering as the durian, but more fragrant than that of a mango. These enormous collective fruits grow by the dozen, sprouting out from flimsy trunks that seem far too feeble to support their hefty weight. Climbing up to harvest them was very easy, and compared to the featureless coconut trees, scaling these branch-ridden boughs seemed like a walk in the park.[31] Furthermore, jackfruit

---

31 A collective fruit is a cluster of smaller fruits fused together, so that it appears as one large fruit. Collective fruits may be identified by multiple seeds each encased in its own aril.

trees were now a common sight in the lowland rainforests of Hainan, having become naturalised after their introduction from the Western Ghats of India over a thousand years ago.

Jackfruit – by far the most delicious, productive and wholesome of all the rainforest fruits

If the fruit alone seems almost too good to be true, one may be surprised to hear that the entire tree is a treasure trove. Jackfruit wood is attractive, durable and termite-resistant, making it one of the best raw materials for making high-end furniture. Within each individual fruit is an ovate brown seed about the size of a new potato, which has a very similar starchy taste and nutty flavour when boiled and peeled. Unfortunately I did not have a pan to roast them with, or else they would have tasted even better, I imagine. Young jackfruits have a mild flavour and stringy internal structure, which apparently makes them taste like pulled pork and have made them a popular meat substitute in vegan diets. Personally, I found this practice rather ludicrous – to cut off an immature fruit and deny its many offspring of survival was, to me at least, no less 'cruel' than harvesting the eggs of hens. Since I was living almost entirely off jackfruit towards the end of my trip, going through an astonishing amount of the stuff every day, I steered well clear of the unripe ones and made sure to leave some of the seeds uncooked so that I could scatter them afterwards – hopefully, this went some way towards ensuring the plant's continued survival in the area.

Jin, however, was not such an avid fan of the jackfruit. He thought the smell was disgusting, and was not best pleased when he found his fingers stuck together by the latex oozing from the fruit's stem and rind. To me, this was only a minor annoyance, and more than made up for by the exquisite taste of the pulp inside – it was truly heavenly, and the boy watched in disbelief as I gobbled down an entire twenty-pounder in the space of an hour. Energised and ready to go, I cut two more ripe fruits down with the machete and tied their stems together with a short length of rope – it was nearly dusk, and these were to be my rations for the last two days of my stay. As I lugged my hefty bounty back onto the road, I turned to see that Jin had wandered off on his own.

'Come on, Jin, it's getting dark!' I yelled. 'Grandma's going to be angry at you if you don't get back on time!' I remembered my promise to the old lady, and realised that the clock was ticking.

I could barely hear myself. My yells were drowned out by the evening chorus of siren cicadas.

'Oh, for goodness' sake!' I laid the jackfruits down on the road, and plunged back into the darkness of the forest.

But I couldn't find Jin. I looked all around the jackfruit trees, the tree where we caught the dragon... I searched every place that we'd passed through during the day, but to no avail.

'Jin!' I screamed at the top of my voice.

The boy seemed to have vanished without a trace.

Where could he have gone? I began to hack my way through the dense jungle and up towards a steep ridge, the other side of which sloped down to join the back of the village. *Maybe he went back by himself? Took a shortcut over the brow of the hill?*

But it was unlike him. Jin always lent me a helping hand — whether it was carrying some of the equipment or sorting through the catch. To disappear without warning was the last thing I could imagine him doing. While I brooded over all the different possibilities, I emerged into a glade just before the steep section of the hill. It was a most peculiar sight.

Decrepit walls and shattered windows

Out from the thickets there sprung a huge concrete structure, whose lifeless form sprawled over the weeds and occupied most of the small clearing. From its central portion extended two wings, and through the doorway I could see nothing but a tangle of plants. The overall lacklustre design suggested a modern origin, although the decrepit walls and shattered windows told of years of abrasion and neglect – the place had clearly been abandoned long ago. Upon the porch hung a wooden sign written in Chinese calligraphy, which I was always hopeless at reading, but the characters seemed to indicate that this was some kind of museum. Just as I was about to cross the glade and continue my search uphill, however, I heard an unsettling noise coming from the ruins.

It was the giggling sound of a child. For a moment I thought I was hallucinating, but then another muffled laugh echoed from inside the building.

'Jin, is that you?' I tried to sound calm, but I could hear my voice trembling.

There was no response.

I clutched the machete, and stepped into the complex.

THE MUSEUM

An open hallway stretched out to either side, which looped around to form a square around the central courtyard. I turned to my right, and saw a figure hunched over a table with his back to me. The last rays of sunset shone through the columns of the peristyle, illuminating a row of large wall cabinets full of taxidermic specimens. Vines, mosses and palms had overwhelmed the place, to the extent that it looked more like a tropical arboretum than a museum. I gave a sigh of relief as the person turned around to reveal a smiling Jin.

'Come look at this, sir!' The boy sprinted towards me and pointed at one of the wall displays, terribly excited.

'Really, Jin! You could have told me before you ran off,' I grumbled as I walked across the hallway. 'I almost had an anxiety attack!'

The boy said nothing and simply smiled back at me, his finger still pointing in mid-air. I got up close, and chuckled at the strange creature in the cabinet.

'Ha! It's a pangolin, Jin. Have you never seen one before?' As I uttered those words, I struggled to recall if I had ever seen a living one myself.

'No, sir. Is it real?' The boy stared in awe at the prehistoric-looking creature. 'What's it doing?'

'Of course it's real! Well, the skin at least – all the insides have been removed and stuffed with cotton wool or something similar.' I gestured towards the fork in the tree. 'And you see that big lump on the trunk? That's an arboreal ant colony; that's what it's after. The long tongue is used for reaching into the narrow tunnels and mopping up the ants. It's probably plastic though; the real one would've rotted away by this point.'

Jin stood in silence, utterly captivated by the pangolin. I could imagine that to a boy who had never visited a zoo, a museum or even watched television, seeing such a curiosity must have been powerful beyond description. After a lengthy observation, Jin finally broke the silence.

'I would love to see a live pangolin, sir. Could we try and find one?' The boy blinked at me expectantly.

'We would never find one, Jin. Lots of people eat them.' I sighed. 'And use their scales for medicine. They're as good as extinct now.'

The boy stared in awe at the prehistoric-looking creature

A look of confusion and disgust swept across his face.

'Ugh!' The boy recoiled, clearly repulsed by the idea of eating such a strange-looking thing. 'Why can't they just leave it alone?'

My heart sank like a stone at these simple words. The boy clearly had more sense than the majority of the population, but there I stood – as an ambassador of the more 'advanced' society who needed to give a logical explanation for our atrocities. There was none.

'It's... self-delusion and social status, Jin. I'm not sure it's something you would understand.' I looked away in shame. 'And I hope you will never have to.'

The boy gazed with melancholy at a leopard cat sitting at the bottom of the cabinet. I could see his eyes tearing up, just as mine were, and I remembered those times when I had wept for a similar cause – for the unimaginable suffering of wildlife, of plants and animals that existed long before our destructive race came into being. No amount of reassurance or consolation could have cleared the intense rage and confusion inside him, and I couldn't bring myself to watch any further. I took a deep breath and held back my tears, leaving the boy behind as I walked off to examine the table display case. The gentle sound of sobbing echoed from the other end of the corridor.

A large entomological collection lined the bottom of the table display case: moths, butterflies and beetles, all superbly curated and in pristine condition thanks to the airtight cabinet. The immaculate specimens formed a strange contrast with the surrounding dilapidation, and exuded a sense of otherworldly beauty. As I scanned through the collection, my gaze was immediately drawn to a pair of familiar-looking beetles in the corner. They were identical to the one that had sprayed me up near the peaks.

My heart began to race. I leaned my head against the glass to read the collection data. They were written in scrawling characters on tiny paper labels, virtually hidden underneath the specimen itself. Even if the cabinet was open, I would have had a hard time deciphering the information – with the cabinet locked and the lighting so dim, I didn't stand a chance.

'Have you seen this one, sir?'

I jumped in surprise. I was so absorbed in my own thoughts that I had forgotten Jin was there. I looked up and saw that he was now at the far end, admiring the display inside one of the wall cabinets. My pupils dilated – I could see the beetle from my end of the hallway.

I rushed over, and gaped at what I could only describe as a monster. A bronze shell encased a huge thickset body, which seemed more fit for a turtle than a beetle; towards the front, a brilliant green pronotum shone like a knight's armour, cast into the most fantastic shape and embellished with fringes of orange hairs. But most extraordinary of all were those arms…

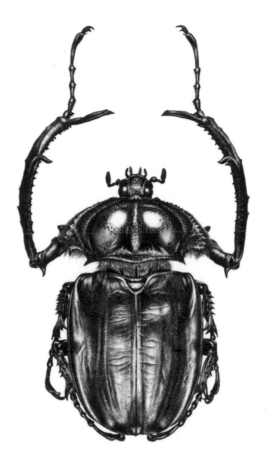

The owner of that mysterious limb

'The arms, sir! They look identical to the one you found!'

Jin had echoed my thoughts. This was the owner of that mysterious limb I had found in the flood debris, there was no doubt about it. The curved tibia, the proliferation of spikes, the extra tarsal claw… it all came flooding back to me. That beautiful evening seemed so long ago. *What will I tell Max?* I leaned back against the wall and sank slowly to the ground.

'What's the matter, sir?'

'I have failed, Jin.' I stared blankly at the bloodstained sky. 'This whole trip has been for nothing.'

'There's still hope, sir.' The boy looked at me sternly. 'The pitfall traps. I'll go check 'em tomorrow.'

'I can't let you do that, it's far too dangerous.'

'Well, I managed to follow you up there, sir.' He grinned mischievously. 'I'm sure I'll be able to do it again.'

'All right, all right,' I conceded grudgingly, but deep down I thanked the boy for the kick of motivation I desperately needed. 'You really are trying at times, Jin!'

JIN'S FAREWELL

The cups lay cracked and strewn across the leaf litter, such that there couldn't have been a greater mess if a herd of bison had come stampeding through. I should have known that nothing edible was safe up in these cloud forests, and some animal must have smelled the bait. My efforts had been completely vandalised – I had failed once and for all.

'I'm sorry, sir,' the boy whimpered, half-expecting me to burst into a rage. 'I shouldn't have brought you up here.'

I stood there staring blankly at the destruction on the forest floor, not quite knowing how I should react. Disappointment had become my norm, and I was now numb from the endless setbacks.

'It's OK, Jin. It's not your fault.' I forced a smile as I gathered up the broken pieces of plastic, and patted him on the shoulder. 'Plus, we had a pleasant hike today, right?'

That part was at least true. For once the thick fog had cleared over the peaks, and the cloud forest was transformed into the most beautiful botanical garden: orchids, bromeliads and ferns hung as ornaments from the twisting trunks, which were covered by cushions of green velvety moss. The boy smiled back at me – a genuine smile of contentment, of a feeling that everything was perfect as it was. He then turned and raced towards the edge of the granite platform.

'Careful, Jin!' I cried with panic, but the boy had everything under control. With an agile shuffle he sat on the platform edge, his feet dangling over the abyss. I walked up slowly, and sat down next to him.

It was a magical sight. The entire mountainside glimmered under the warm glow of the afternoon sun, lighting up the distant rolling hills for as far as the eye could see. I took in a long, deep breath and, for a moment, the absurdity of my adventure seemed to vanish into the crisp mountain air. We both sat there as still as statues, watching the pure clouds shape-shifting through the blue, and letting them carry our thoughts far away into the distance.

'When are you leaving, sir?' Jin said, after a lengthy silence.

'Tomorrow morning. My driver's picking me up at seven,' I replied slowly. 'Before I go, there are some things I need to…'

'Will you ever come back, sir?' There was a look of sadness in the boy's sparkling eyes.

'I will be back, Jin.' I lifted up his hand and gave him a pinkie promise. 'Next year at this time, when the monsoons return. You have my word.'

The boy nodded.

'There is a specimen case in the cupboard, with a few of your catches pinned up already. That's for you to keep, and I've left some insect pins in there as well for you to expand on the collection.' I disassembled the butterfly net as I spoke, and packed it into a plastic bag. 'And this – it's now yours.'

Jin stared at me, wide-eyed in amazement. 'Are you sure, sir?'

'Of course! This is your prize!' I gave him a cheeky wink. 'You caught the dragon, remember?'

'Thank you, sir!' The boy took the net, beaming with delight. 'What should I do with the insects I catch?'

'Oh, you can do whatever, Jin. Play around with them, observe them, let them go...' I put up a finger in the air and gazed at him sternly. 'But only kill them when it's absolutely necessary, OK?'

'Understood, sir! How can I preserve them without the spirit jar though?'

That was a real poser.

'Hmm... I think I'll take out my resealable packets, and decant the ethanol back into the jar. It's only for a day or two, and I'll take them to the museum as soon as I get back to England,' I said. 'Besides, I don't think the customs officer would be best pleased with a huge jar of alcohol, not least one floating with beetles! But let us hurry – time is running out and we need to get off this mountain before nightfall.'

I helped Jin up to his feet, and we hobbled down the labyrinth back towards the village.

THE GREAT
ENTOMOLOGIST

Cambridge, June 2017

As my fellow graduates gathered on the lawn of the Senate House and chattered away over their champagne, I rushed back to my room incensed by the vanity of the whole affair. I could have stayed in the jungle hunting *Carabus* amongst the cloud forests, but instead I was standing here in a bow tie and gown, celebrating three years of uninspired bookwork and empty theory-spinning. As quick as I could I changed into ordinary clothes, grabbed my packets of specimens and hopped on the next express train to London.

Within an hour or so I found myself racing through the grand terracotta archways of the museum, headed right for the internal collections. This was the moment of truth I had been waiting for.

'Max, I'm back!' I declared as I threw open the office door.

The whole place was silent. The lights were still on, but all six workspaces were empty.

'Max? Michael?'

It seemed as if the whole Coleoptera department had disappeared into thin air.

I went out onto the glass walkway. The other departments – botany, palaeontology and even the rest of entomology – were all busying away as usual. What had happened? As I stood confused in the corridor, I saw a familiar face emerge from the Hymenoptera department.

'Gavin!' I waved at him excitedly, and remembered the specimen I had saved for him. 'I've got a present for you!'

'Hello, my friend! You look like you've been on some adventures recently!' he exclaimed – I was brown and thin from my travels and I think my appearance must have shocked him. He shook my hand with enthusiasm. 'Not a parasitoid again, is it? I still remember that *Paraperithous gnathaulax* which you reared from *Pyrochroa* all those years ago – still one of the most impressive specimens in our British Ichneumonid collection!'

'Well, I'm afraid I haven't moved on from there!' I chuckled, and handed him the packet containing the little wasp. 'The host was this strange globular creature I couldn't identify – I have some photographs here if you…'

'Ha! It was a Limacodid caterpillar then, for sure!' He didn't even look up, and continued to examine the tiny insect. 'Because this is a Braconid wasp, and they specialise in parasitising Limacodids. As for the specifics… this is mostly likely a *Spinaria*, although I couldn't tell you the exact species off the top of my head – it's a very obscure genus, you see?' The expert laughed the revelation off as if it were nothing.

I stood there, quite speechless, trying to grasp how on earth he had processed that much information in the space of thirty seconds.

'Well, that's… quite remarkable!' I stammered.

He smiled back at me. 'Well listen, thank you so much for the specimen, but I must dash off now. Anything else?'

'Yes – one question,' I suddenly remembered. 'Where have all the coleopterists gone?'

'Haven't you heard? They're in the process of moving the entire Coleoptera collection.' Gavin began jogging backwards as he spoke. 'It's still sitting in that old Victorian building over there. All the coleopterists are running back and forth like headless chickens at the moment.'

'Great, thank you! And let me know when you have the specifics on that wasp!'

The scientist gave me a thumbs up as he disappeared down the corridor.

After a lengthy detour, I eventually found the old Victorian building that housed the beetle collections. The place was dark, damp and musty – a far cry from the modern glass construction of the museum's western wing. I took a deep breath, and stepped inside the storage room.

Max was busy fixing up some specimens that had been broken in the relocation, and I waited patiently as he finished this meticulous task. When he eventually looked up, I handed the specimens to him and began to ramble incessantly about my adventures – about how I chanced upon the light trap, about my turbulent trips to the peaks and about my failure to catch *Carabus*. He listened intently to all this as he studied the specimens in the packets, and proceeded to ask me some truly bizarre questions which made me question his sanity. It was only when he pointed out that Mêdog was a restricted military zone and inaccessible to foreigners that I began to see what he was hinting towards.

'Are you implying, then, that Hakomoto never actually collected these himself?' I asked. 'I once suspected that as well, especially after finding that light trap deposited in the Hlai village. But here's the catch – nobody ever went back for those specimens.'

'No, no!' Max chuckled at my simple-mindedness. 'Hiring local collectors – a classic but unreliable method. The species from this collection are extremely sporadic in occurrence, and even the most accomplished collector would not be guaranteed to find these year after year. Yet this man has sought you out to promote his "prized trophies" for future sales, and manages to stock hundreds, if not thousands of them a year.'

'Now that you put it that way, it is indeed most perplexing.' I nodded in agreement.

Max smiled and raised his eyebrows like a mischievous kid. He seemed to be deriving much entertainment from my slow-wittedness.

'Have you ever contemplated the possibility that none of this is actually real?'

'What do you mean, *not real*?' I retorted. The entomologist's vague questions were really trying my patience. 'I'm genuinely beginning to think that you're as clueless as I am, Max.'

The trick worked. Max leaned forwards in his chair with an air of concentration, and placed his hands together on his chin.

'All right, all right. Let's start with this.' He glanced at the tattered specimen box lying on the table. 'Flimsy cardboard, with no return address. Is that how you would post a collection worth several hundred pounds?'

'No, of course not.'

'Good, so now we know that in Hakomoto's mind, this

collection is worth a great deal less than he makes it out to be. Why? Because he can churn them out like a production line. But he knew you would agree to keep it as your pay because, in your mind, it is more valuable than money.'

'Yes. I mainly kept it because I knew Mêdog was a famous holotype locality for *Lucanus,* and as far as I knew *Carabus* had never been recorded from Hainan before – so there was a high likelihood that they were new species.'

'Exactly! He knew you would notice these details.' Max's eyes lit up with an intense sparkle. 'So what's to stop him from faking the collection data?'

'Nothing,' I admitted. 'But I'm sure he's been to some of those places – I saw his machete marks on a dead tree five thousand feet up that mountain.'

'Only to record fake locality data and source some of his breeding pairs.'

'Breeding pairs?' I echoed. I was becoming fascinated by the twists and turns of this profound mystery.

'Yes. Even the specimens aren't real, I'm afraid. They are Hakomoto's creations,' Max continued. 'All captive bred under controlled laboratory conditions, and that's why they're absolutely flawless. The purple variant of *Rhomborhina* is the product of many generations of selective breeding, and I ran some molecular tests on the *Carabus* specimen – it's a hybrid! Between the species *Carabus ignimitella* and *Carabus pustulifer* – as you correctly observed it had similarities to both. All of the species I just mentioned are common on the mainland, and that's probably where the lineages came from. As for the *Lucanus* from Tibet… now that is more interesting. He probably obtained the source from just over the border, either Assam in India or Kachin in Myanmar. Have you heard of *kinshi-bin?*'

'Yes, I believe I have.' I was relieved that I finally knew something. 'Is that the fungus-infused sawdust invented by Japanese beetle breeders?'

'Exactly!' Max was now glowing with excitement. 'Full of protein-rich mycelium, and perfect for the slow-growing *Lucanus* – shortens their larval development by months, and produces adult males with monstrous head capsules. Almost like beetle steroids, if you like!'

'This is brilliant, Max!' I shook my head in astonishment. 'But what was all that business about longhorn beetles and living plants? I still don't understand!'

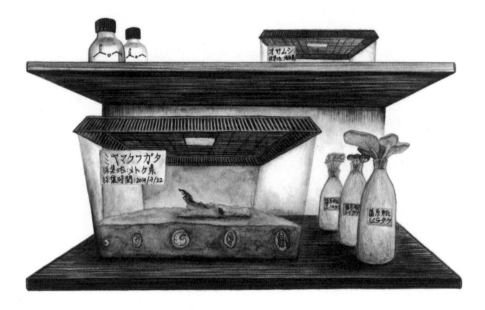

Hakomoto's creations

'*Think about it!*' Max suddenly frowned and pointed a finger up in the air. 'How would you rear something that develops inside a living tree, especially when all the species have different host preferences? Take this *Dorysthenes*, for example,' he lifted up the specimen that he had just repaired, 'whose larvae develop inside the roots of sugarcane, coconut and other monocotyledons. Cross over to the related genus *Prionus* and they tend to prefer the roots of hardwoods. Trying to breed these would be like running a botanical and zoological garden simultaneously – a nightmare to sustain! All of these species have been carefully chosen to be low-maintenance, the sort which you could leave in a pile of debris at the bottom of a fish tank and come back to in a year to reap the rewards. Should Hakomoto, if that's even his real name, have gone about breeding longhorn beetles, he would have needed a whole lot more than a fish tank to satisfy their diverse requirements. Clear?'

'It's as clear as daylight.' I sighed. 'What an extraordinary illusion! But when did you figure all of this out, Max?'

'I formed my hypothesis as soon as I saw the collection.' His frown relaxed into a gentle smile. 'And the theory gradually matured as I mulled over the facts many times, while you were probably halfway up that mountain. When you told me about the signs that Hakomoto had left behind and your failure to find *Carabus*, my suspicions materialised into an inevitable conclusion. Please forgive me for not telling you immediately, as I did not want your search to become biased by my own opinions,' Max explained as he saw me look at him blamefully. 'Besides, who knows? *Carabus* might still be up there – under some unturned stone in the cloud forests of Hainan, still waiting to be discovered.'

I sank into silence. I had been chasing an illusion all this time, oblivious to the glaringly obvious clues that gave away the collection's true nature. The sheer extent of my failure finally hit home.

'I should probably go, Max. I'm keeping you from your work.' I got up to leave, but suddenly felt a funny object in my pocket.

'Oh, just before I go – what do you reckon this is?' I took out the giant limb, and handed it over to him.

'How remarkable!' he muttered as he scrutinised the curiosity. 'Tarsal structure is distinctly scarabaeid – but the sheer size! Almost reminds me of the Neotropical dynastines, but that can't be possible…'

I couldn't help but feel a sense of triumph as the great entomologist looked puzzled for once. *This is how it feels to be stumped, Max.* I gave a half-smile. *I have the upper hand this time.* But my victory was short-lived – within seconds the wrinkles on his forehead evened out, and he burst into a hearty laugh.

'How foolish of me, it's… it's a Euchirine! Genus *Cheirotonus*, to be more precise. No, wait – it's *Cheirotonus jansoni*! I'm willing to bet my life on it.'

'How could you possibly be so sure of that?' I gasped in awe. *Cheirotonus jansoni* was indeed the name under that specimen in the abandoned museum.

'Experience, my friend!' Max grinned back at me, proud as a magician who had just received applause from his audience. 'When all you see every day are beetles, you start developing a feel for them – even if that means learning four hundred thousand different Latin names!'

METAMORPHOSIS

Hainan, June 2018

'Didn't you hear me? *This one's full!*' the driver yelled impatiently. 'You'll have to wait an hour for the next one.'

I blinked, and shielded my eyes from the glare of the afternoon sun. The minibus rolled away, sputtering out vile fumes from its exhaust and leaving a cloud of dust in its wake. I coughed as I got caught up in the whirlwind, and crossed the road to the other side. There was at least some greenery here, which soothed the eyes from the porcelain exteriors of this garish town.

After a year, I was back to fulfil my promise. *I wonder what Jin has been up to.* I ambled up and down the pavement, lost in my own thoughts.

Maybe he's been up to the peaks again, and found Carabus *already. Or maybe* Carabus *simply doesn't exist, and we're doomed to fail again.*

I wandered on aimlessly along the road, tapping at the shrubs and weeds with my new net. The time passed quickly – soon I had lost sight of the bus station, and found myself at a busy intersection next to an open-air market.

My gaze immediately fixed onto an unusual shape dangling from a creeper vine, and my heart began to pound violently.

What is that?

It was a majestic birdwing butterfly in the process of emerging from its chrysalis. This would have been a rare discovery even in the jungle, let alone in a place like this.

As the butterfly continued its mesmerising transformation, I was flooded with a sense of awe. Right here before me was an event that I had encountered countless times in the past, but never took the time to appreciate its significance – like a phoenix rising from the ashes, the butterfly had reassembled itself from the dissolved remains of its caterpillar past. How the insect managed to do so with such precision and control was, surely, beyond anything that we could hope to understand. Yet here I was – standing in the middle of a busy street, the only witness to this spectacle. Some passers-by gave me weird looks, others groaned as they diverted around me, but most of them were simply tapping away at screens, completely oblivious to their surroundings.

For a moment I felt an intense surge of pity. *What is this poor creature doing here?* Every gust of dusty wind from the passing cars threatened to tear it into pieces, yet its fragile claws somehow clung on. I could not imagine what food or shelter it would find in such a forbidding place, let alone a mate. *What a futile struggle!* I sighed. Then, as if it could sense my concerns, the butterfly flapped its wings and took off into the breeze.

I was struck with a most ineffable feeling, which I could only describe as a *satori* experience – a feeling of complete calmness

A majestic birdwing butterfly

and liberation.[32] Here I was, mourning for the butterfly's inevitable demise, but to the flow of time is my own life not as fleeting as his? Perhaps he has already seen through the void of existence, and realised that his dreams and desires are nothing more than drifting clouds that shape-shift across the sky. Indeed, an illusion may await him at the end of his journey, but it is the process that he cherishes, along with the memories that he will leave behind. *He knows the challenges he will face. He knows that the clock is ticking.* That is precisely why he has taken this treacherous path. *He is a pioneer for his species, an adventurer into the world of the unknown.* It may well be that few will ever notice him, but those who do he will captivate and delight. It may well be that some callous person will trample him into the ground,

---

32 *Satori* is a form of enlightenment, commonly used in Zen Buddhism to describe a deep spiritual experience that allows one to 'see into one's true nature'.

but he will keep fighting until his last breath. Or maybe he will enlighten an observant soul, who realises that beauty and wonder are everywhere if we are mindful. And so, he flies on with his undying dreams, golden wings glittering under the evening light.

ACKNOWLEDGEMENTS

I would now like to take this opportunity to thank some special people, without whom the completion of this work could not have been possible. First and foremost is the great entomologist Max Barclay, whose poignant foreword stresses the importance of natural exploration in the modern age. He, along with Michael Geiser and Keita Matsumoto from the Coleoptera department, have been my friends for many years and our interesting conversations provided endless inspiration for the tales in this book. I am also very indebted to Gavin Broad, who was the hymenopterist detective on the case when I brought my first interesting specimen to the Natural History Museum – I hope he likes his cameo appearance!

On the artistic front, I am honoured to have had the opportunity of collaborating with Carim Nahaboo – one of my biggest inspirations and by far the most prolific artist I have ever met. He provided the fantastic pen-and-ink vignettes for each chapter, all lavishly illustrated with detail which together form a huge asset to this book. None of my own endeavours could have been possible, however, without the guidance of the late Terry Vincent and the teaching from Gabriel Forbes-Sempill, who introduced me to the wonderful world of art and remains one of my closest friends. Many of the illustrations in this book made use of photographic references, and I must thank the brilliant macro-photographer John Horstman for agreeing to let me use his images for painting reference. His intriguing images and records about Limacodid caterpillars provided inspiration for one of the narrative strands, and I sincerely hope he continues his fascinating work in Pu'er, Yunnan.

As the storyline was fairly complex, I was very keen to get some outside opinions before I finished the final draft. I would like to thank my good friends Katherine Alcock, Tim Ekeh, Ben Herd and Ashton Sheriff for being my beta readers and checking for any inconsistencies. And last but not least, I would like to thank my family for their kind understanding over the years, and for providing the financial support for all my adventures and endeavours – my grandparents Yongfu Sang and Yongrong Zheng took me on many bug hunts when I was a child, and it is arguably them that I owe the most for my love of nature.

A NOTE ON THE AUTHOR

Yikai Zhang is an artist, writer and entomologist with a degree in Natural Sciences from the University of Cambridge. He is currently studying biodiversity at the Natural History Museum in London, where he has been a volunteer curator for many years.

Unbound is the world's first crowdfunding publisher, established in 2011.

We believe that wonderful things can happen when you clear a path for people who share a passion. That's why we've built a platform that brings together readers and authors to crowdfund books they believe in — and give fresh ideas that don't fit the traditional mould the chance they deserve.

This book is in your hands because readers made it possible. Everyone who pledged their support is listed below. Join them by visiting unbound.com and supporting a book today.

Kate Abernethy

Danny Adhami

Darius Afkhami

Ransford Agyare-Kwabi

Alen Akhabaev

Hasan Al-Rashid

Katie Alcock

Stephanie Andrews

Mercy Bannister

Max Barclay

Paul Bentley

Lawrence Berry

Sebastian Berry

Xueni Bian

Stefan Blaser

Harry E. Blevins, Jr.

Yang Bo

William Bond

David Bradbury

George Brill

Armand Brochard

Jack Brodsky

Martin Brown

Alice Cao

Chloe Casey

Daniel Chao

Charlie Chen

Huan Chen

Lei Chen

Mo Chen

Peiliang Chen

Wei Chen

William Chen

Kang Cheng

Xiaotong Cheng

Xinyu Cheng

Demos Christou

Christy Chung

Michael Chwu

Ben Clowes

Neil Clowes

Alex Cranston

Haoyu Cui

Yongyan Cui

Junchen Dai

Junqing Dai

Victoria Samantha Dawson

Jaime de la Torre

Amala Desai

Anya Doherty

Shengjie Dou

Linda Doughty

Wanzu Du

Trisha DuCharme

Connor Duffy

Elizabeth Dunne

Jill Eaton

Tim Ekeh

Pauline Elliott

Jane Evans

Ben F-N

Jinglu Fan

Zehua Fan

Jacky Fang

Mark Feltham

Lizzy Fone

Diyun Fu

Sunny Fu

Stephen Gaillard

Mark Gamble

Rosie Gangar

Mingyang Gao

Stephanie Gao

Michael Geiser

Lily Goekjian

Fengyuan Gong

Yi Gong

Neirin Gray Desai

Shan Hai

Archie Hall

Ian Hammond

Gloria Han

Yanhua Han

Thomas Hanton

David Harford

Paul Hartley

Thomas Hartley

Anne Henderson

Ben Herd

Tom Hill

David Ho

Lijie Hong

Mike Hong

John Horstman

Shaun Houlden

Bob Howell

Chen Hu

Han Hu

Zixuan Hu

Changcheng Huang

Ying Huang

Andreas Ioannou

Ceri James

Jakob Jazbec

Qiaomei Jiang

Xingyi Jiang

Yanzhi Jin

Alex Jones

Samuel Jornert

JSM

Dan Kieran

Andrew Kirby

Yuma Kitahara

Harvey Klyne

Chilla Knight

Julia Koenig

Chris Kwok

Garth Leder

Joshua Lee

Eliza Levien

Bin Li

Cindy Li

Jiashu Li

Jing Li

Nan Li

Ryan Li

Tianze Li

Wenwen Li

Zhuofan Li

Rachel Liao

Jie Lin

Li Lin

Cathy Liu

Guanghui Liu

Hui Liu

Kunyu Liu

Shiqi Liu

Yongzhi Liu

Sharon Low

Shu Lu

Qiyi Lü

Yanqiao Lü

Peixi Luan

Mike Lynd

Felicia F. Martinez

Kenki Matsumoto

Wei Meng

Benjamin Merrett

John Mitchinson

Lisa Mldsova

Fernando Mosler

Charlotte Mounter

Carim Nahaboo

Carlo Navato

Thomas Nightingale

Yu Okamura

Weihong Pan

Will Pannetier

Raghul Parthipan

Zhong Pei

Chuyue Peng

Justin Pollard

Meng Qi

Jiayi Qiao

Hugo Ramambason

Sushila Ramani

Clare Rees-Zimmerman

Faye Robertson

Edward Rong

Kalina Rose

Abhilash Sarhadi

Chiara Schaefer

Francisca Sconce

Yoel Sevi

Theodore Shack

Nikhil Shah

Jing Shao

Mingwei Shen

Wei Shen

Bin Shi

Shi Shi

Odysseas Sierepeklis

Pawat Silawattakun

Fiona Silk

Demetris Skottis

Yi Song

William Stevens

Alistair Stewart

Philip Stewart

Jundong Tang

Angelo Tata

Oliver Tesh

Jessa Thurman

Zhuo Tian

Mira Manini Tiwari

Meg Tong

Ann Trevorrow

Oskar Ulvestad

Kent Vainio

Charles Vallee

Shariq Varawalla

Linda Verstraten & Pyter Wagenaar

Terry Vincent

Jenni Visuri

Wa Wa

Sir Harold Walker

Jiankun Wang

Jingyuan Wang

Qi Wang

Shenlei Wang

Wei Wang

Wen Wang

Yonggang Wang

Guofu Wei

Westminster School Library

Daisy Wheller

Gavin Wilson

Jakob Wolf

David Wong

Jun Wu

Rutong Xia

Feng Xiao

Hairong Xie

Kenie Xu

Pingfei Xu

Sanqiao Xu

Xiaogang Xu

Xiaokang Xu

Yaya Yan

Hao Yang

Kathy Yang

Lin Yang

Shaonan Yang

Xingying Yang

Yingshuang Yang

Zhuo Yang

Peng Yao

Ping Yao

Stan Yao

Xiaoyao Yao

Kai Yu

Xiang Yu

Xinge Yu

Bing Yue

David Yue

Jessica Yung

Baoxia Zhang

Chi Zhang

Hengyan Zhang

Jian Zhang

Jianxin Zhang

Shunhong Zhang

Siteng Zhang

Steven Zhang

Wanquan Zhang

Xianpu Zhang

Xiaokang Zhang

Xubo Zhang

Yikai Zhang

Yujuan Zhang

Zhiping Zhang

Pingping Zhao

Guitong Zheng

Yang Zheng

Yuhua Zheng

Min Zhou

Fangzhou Zhu